Statics
FOR
DUMMIES®

by James H. Allen III, PE, PhD

WILEY

Wiley Publishing, Inc.

Statics For Dummies®

Published by
Wiley Publishing, Inc.
111 River St.
Hoboken, NJ 07030-5774
www.wiley.com

For general information on our other products and services, please contact our Customer Care Department within the U.S. at 877-762-2974, outside the U.S. at 317-572-3993, or fax 317-572-4002.

For technical support, please visit www.wiley.com/techsupport.

Wiley also publishes its books in a variety of electronic formats. Some content that appears in print may not be available in electronic books.

Library of Congress Control Number: 2010930963

ISBN: 978-0-470-59894-8

Manufactured in the United States of America

10 9 8 7 6 5 4 3 2 1

WILEY

About the Author

James H. Allen III, PE, PhD, serves on the civil engineering faculty at the University of Evansville, where he teaches statics, structural analysis, and structural design courses. Dr. Allen received his PhD from the University of Cincinnati in structural engineering and performed his undergraduate work at the University of Missouri-Rolla (now the Missouri University of Science and Technology).

Dedication

To my wife Miranda for her unconditional love and support.

Author's Acknowledgments

I wish to thank all of the many people that have worked so hard to make this book a reality. Thanks especially to my senior project editor, Alissa Schwipps, for her patience and guidance, and to all of the others who have made this project possible, including Mike Baker, Megan Knoll, and Wiley's Composition Services department. Thanks for all you do!

Publisher's Acknowledgments

We're proud of this book; please send us your comments at `http://dummies.custhelp.com`. For other comments, please contact our Customer Care Department within the U.S. at 877-762-2974, outside the U.S. at 317-572-3993, or fax 317-572-4002.

Some of the people who helped bring this book to market include the following:

Acquisitions, Editorial, and Media Development

Senior Project Editor: Alissa Schwipps

Acquisitions Editor: Mike Baker

Copy Editor: Megan Knoll

Assistant Editor: Erin Calligan Mooney

Senior Editorial Assistant: David Lutton

Technical Editors: Joshua Hertz, Valery N. Bliznyuk

Senior Editorial Manager: Jennifer Ehrlich

Editorial Assistants: Jennette ElNaggar, Rachelle Amick

Cover Photos: © Getty Images

Cartoons: Rich Tennant (`www.the5thwave.com`)

Composition Services

Project Coordinator: Patrick Redmond

Layout and Graphics: Nikki Gately, Kelly Kijovsky, Melissa K. Smith, Erin Zeltner

Proofreaders: Laura Albert, Henry Lazarek

Indexer: Rebecca Salerno

Special Help: Elizabeth Kuball

Publishing and Editorial for Consumer Dummies

> **Diane Graves Steele,** Vice President and Publisher, Consumer Dummies

> **Kristin Ferguson-Wagstaffe,** Product Development Director, Consumer Dummies

> **Ensley Eikenburg,** Associate Publisher, Travel

> **Kelly Regan,** Editorial Director, Travel

Publishing for Technology Dummies

> **Andy Cummings,** Vice President and Publisher, Dummies Technology/General User

Composition Services

> **Debbie Stailey,** Director of Composition Services

Contents at a Glance

Table of Contents

Introduction

● ●

*A*s I watch students working toward mastering the principles of statics, I find myself frequently answering some of the same basic questions. Despite countless hours of working through examples and homework problems from their textbooks, students often seem to be confused on the same several topics.

The problem isn't that the material in a typical statics class is overly difficult; I think the issue is just several simple misconceptions that manifest themselves through poorly written examples and unnecessarily complex wording in conventional statics textbooks.

That's why I've written *Statics For Dummies* — to help students of the subject get a better understanding than they may otherwise get in a classic textbook. In this book, my goal is to answer those basic questions by using simple explanations and eliminating a lot of the extra technical jargon.

About This Book

No statics book can tell you how to solve every possible problem you encounter. What *Statics For Dummies* tells you is what you need to know and why you need to know it. Why are three-dimensional problems easier to solve with vector formulations than with scalar methods? What exactly is equilibrium, and how do Newton's laws guarantee it? How do you know the difference between a truss and frame? All of these topics are at the heart of understanding statics; after you've got these basics down, actually solving a statics problem is a snap!

In statics, one of the most important habits to form is being as methodical as possible, which means that statics lends itself very nicely to a large number of checklists or simple steps to remember and follow. Throughout this book, I try to organize certain techniques by outlining the steps that you need to follow. Just like when you go grocery shopping, the checklists help you remember what fruits and vegetables (or equations or free-body diagrams) you need to put in your basket.

The best part of this book is that *you* have complete control on where you want to start. If you just want the tips for solving specific problems, jump to Part VI. If you find you need a bit of a refresher on vectors, that's in Part II. Let the table of contents and index be your guides.

Conventions Used in This Book

I use the following conventions throughout the text to make things consistent and easy to understand:

✔ New terms appear in *italic* and are closely followed by an easy-to-understand definition.

✔ **Bold** is used to highlight the action parts of numbered steps, as well as keywords in bulleted lists.

I also use other, statics-specific conventions that I may not explain every time, so following is a brief list of concepts and terms that I use frequently throughout the book.

✔ **Decimal places:** I try to carry at least three decimal places in all my calculations in this book. This move helps ensure enough precision in my calculations to demonstrate the fundamental principles without getting bogged down in the pesky numerical accuracy issues I cover in Chapter 2.

✔ **Vector variables:** The most important aspect of statics is that you take all effects into consideration; if you forget even the smallest behavior on an object, solutions in statics can become impossible to accurately calculate. To help keep track, I usually use **F** or **P** to indicate force vectors, and **M** to indicate a moment vector. I also use **i, j,** and **k** to represent those common unit vectors in the text; in equations, they appear as $\hat{i}, \hat{j}, \hat{k}$.

✔ **Bold (not in steps):** Aside from its use in numbered steps and bulleted lists, I also use **bold** text to represent a vector equation. If you see a bolded variable, that indicates a vector is lurking in the discussion. This convention is common to most classical textbooks, so I replicate it here just for the sake of consistency with vectors you may have already been exposed to in a conventional statics or physic class.

✔ **Arrows on top of vector names:** Another method of denoting a vector is to use the label or name of the vector with an arrow over the top such as $\vec{F}$ or $\overrightarrow{\text{Weight}}$. If you see an arrow on top of a letter or word in an equation, you know that I'm working with vectors.

✔ **Italics (not as definitions):** I also adopt a second sign convention from other textbooks: When I talk about a vector's *magnitude* (length) in the text, I use the name or label of the vector in italics.

✔ **Absolute value brackets:** To represent the magnitude of a vector in an equation, I surround it with absolute value brackets, such as $|\vec{F}|$ or $|\overrightarrow{\text{Weight}}|$. Because magnitudes are properties of vectors, I still include the vector arrow over the label. Just remember that the absolute value brackets take precedence, so if you see those, you know I'm primarily talking about a scalar magnitude.

✔ **Plus signs (+) with vector senses:** Although it's not required, I use the plus symbol before positive numbers in some vector calculations as a visual reminder that I have in fact considered the *sense* (direction) of the vector on the Cartesian plane.

✔ **Origin:** I assume that the origin of any given Cartesian plane is (0,0) for two-dimensional problems and (0,0,0) for three-dimensional ones unless otherwise noted.

What You're Not to Read

Although I hope you're interested in every word I've painstakingly inscribed in this book, I admit that there are a few things you can skip over if you're short on time or just after the most important and practical stuff.

✔ **Text in sidebars:** *Sidebars* are the shaded boxes that provide extra information, such as history or trivia, about the chapter topic.

✔ **Anything with a Technical Stuff icon:** The in-depth info tagged by this icon is useful, but it may not be entirely necessary to solve day-to-day problems. It may also include information that shows how the information being discussed was developed or how the formulations came about.

Foolish Assumptions

As I wrote this book, I made a few assumptions about you, my beloved reader.

✔ You're any college student taking an engineering statics class or studying Newtonian mechanics in your physics classes, or are at least familiar with those basic concepts.

✔ You remember some math skills, including algebra and basic calculus topics such as differentiation and simple integration.

✔ You have proficiency in geometry and trigonometry. The basic rules governing sines, cosines, and tangents of angles (both in degrees and radians) prove invaluable as you work a statics problem.

✔ You're willing to practice the techniques that I show you. Remember all those problems your math teachers made you work back in school? Statics may require a similar effort. Practice makes perfect!

How This Book Is Organized

This book starts with a basic review of units and math and goes through vectors, forces, free-body diagrams, equilibrium, and practical statics applications. Here's the lowdown on each part.

Part 1: Setting the Stage for Statics

In Part I, I give you some basic refresher information, such as working with units, while reviewing some of the basic math that provides the foundation for statics. Chapter 1 introduces the concept of statics while Chapter 2 provides you with a brief refresher about a wide range of mathematics topics, including basic algebra and polynomials, trigonometric relationships, and basic integration and differentiation of polynomials. Chapter 3 highlights the two major systems of units that you encounter in statics and shows you the base units for a wide range of values in statics.

Part 11: Your Statics Foundation: Vector Basics

Part II introduces some basic vector principles. Chapter 4 shows you the three basic characteristics of vectors and how you can depict them graphically. Chapter 5 describes how to define your first vector, describing direction from one point to another. I also show you several alternative ways to define direction by using vectors. In Chapter 6, I explain the basics of vector mathematics and explore several useful identity relationships that come in handy. Chapter 7 demonstrates how to create one vector from multiple other vectors. I explain several basic techniques and show you how to apply basic formulas for calculations of each technique. Chapter 8 shows you the opposite information from Chapter 7: how you can split a single vector into multiple vectors acting in different directions.

Part 111: Forces and Moments as Vectors

In Part III, I explore how load effects are created. In Chapter 9, I illustrate single concentrated loads (or point loads) and introduce you to the concept of self weight as a single value. Chapter 10 covers loads acting over an area or a distance and shows you how to turn a distributed load into an equivalent concentrated load as well. In Chapter 11, I show you how to calculate the different *centroids* (geometric centers, such as center of area and center of mass/gravity) of different geometric shapes, which proves useful for helping you to locate the single equivalent force of a distributed load. Chapter 12 is where I introduce rotational effects known as moments, explaining how to draw and calculate them.

Part IV: A Picture Is Worth a Thousand Words (Or At Least a Few Equations): Free-Body Diagrams

Part IV shows you how to draw the pictures necessary to solve statics problems. In Chapter 13, I give you the basic checklist of items to include on a free-body diagram (F.B.D.) and then explain how to define supports in terms of forces and moments. Chapter 14 shows you what to draw and how to work with multiple free-body diagrams at the same time. In Chapter 15, I give you some guidance on several ways to simplify some of the more complex diagrams that you create.

Part V: A Question of Balance: Equilibrium

In Part V, I introduce you to the concept of stability or equilibrium in statics. Chapter 16 defines the different types of equilibrium by explaining Newton's three laws of motion and the basic assumptions behind the governing equations of statics. In Chapters 17 and 18, I show you how to apply the basic equations of equilibrium to solve for unknown support reactions in two- and three-dimensional problems, respectively.

Part VI: Statics in Action

In Part VI, I show you how to identify the major categories of problems you come across in the real world. I also highlight several tips and techniques to speed up your solution process. Chapter 19 introduces you to trusses and simple axial members. I show you the basic techniques for solving for internal forces in the members of the trusses. Chapter 20 shows you that for many members in statics, additional internal forces exist beyond just the simple axial cases. I show you how to write equations for these internal forces and how to draw a graph of their values. In Chapter 21, you discover how to deal with frames and machine structures. Chapter 22 provides you with tools necessary to solve for internal forces of systems whose internal forces vary with geometry; I explain the concepts of sag and tension and then provide a useful shortcut technique known as the beam analogy.

In Chapter 23, you sink to new depths by exploring the behavior of fluids on submerged surfaces. I explain the concept of pressure and unit width in your calculations, and how to apply the equations of equilibrium. Things really heat up in Chapter 24 as I introduce friction to the problems. I explain the logic needed to determine if an object is prone to tipping or sliding and how to mathematically prove that.

Part VII: The Part of Tens

Part VII includes helpful top-ten lists. Chapter 25 provides you with ten important statics ideas to remember even if you forget everything else, and Chapter 26 gives you ten tips to survive a statics exam.

Icons Used in This Book

To make this book easier to read and simpler to use, I include some icons that can help you find and identify key ideas and information.

This icon appears whenever an idea or item can save you time or simplify your statics experience.

Any time you see this icon, you know the information that follows points out a key idea or concept, greatly increasing the number of statics tools you have at your disposal.

This icon flags information that highlights dangers to your solution technique, or a common misstep that statics practitioners make but you should avoid.

This icon appears next to information that's interesting but not essential. Don't be afraid to skip these paragraphs.

Where to Go from Here

You can use *Statics For Dummies* either as a supplement to a course you're currently taking or as a stand-alone volume for understanding the basic concepts of statics.

If you're taking a statics course or studying Newtonian mechanics in physics, hopefully you find the organization to be very familiar. I follow the basic topics sequence that you experience in a class. However, unlike a classical text, if you want to skip a chapter, feel free.

If you're studying on your own or have never had a statics class, I strongly urge you to start at the beginning with Chapter 1 and read the chapters in order. The techniques in the later chapters do build on concepts of early chapters. That being said, this book isn't a mystery novel. If you want to skip ahead to the topics at the end, go right ahead; you won't ruin the ending. And if you get really lost, you can always fall back to an earlier chapter for a quick refresher!

Part I
Setting the Stage for Statics

The 5th Wave   By Rich Tennant

"This guy writes an equation for over 20 minutes, and he has the nerve to say, 'Voilà'?"

In this part . . .

This first part introduces you to the basic concepts of mechanics in engineering and the sciences, as well as the differences between the basic fields. You also pick up the basic assumptions you need when working statics problems. I explain the two primary systems of units (U.S. customary and metric) and introduce you to the base units of each system. Finally, I review basic algebra, geometry, trigonometry, and calculus, all of which you encounter frequently in statics.

Chapter 1

Using Statics to Describe the World around You

Statics is a branch of physics that is especially useful in the fields of engineering and science. Although general physics may describe all the actions around you, from the waving of leaves on a tree to the reflection of light on a pond, the field of statics is much more specific.

In fact, statics is actually a part of most physics courses. So if you've ever taken a high school or college physics course, chances are that some of the information in this book may seem vaguely familiar. For example, one of the first areas you study in physics is often Newtonian mechanics, which is basically statics and dynamics.

Physics classes typically cover a wide range of topics, basically because physics has a wide range of applications. Conversely, a statics course is much more focused (which doesn't necessarily mean it's simple). Whoever said that the devil is in the details may well have been talking about statics.

Before you panic, close the book, and begin questioning why you ever thought you could understand statics, let me reassure you that just because statics isn't always simple doesn't mean it's always difficult. If anything, statics does happen to be very methodical. If you follow some basic steps and apply some basic ideas and theory, statics actually can become a very straightforward application process.

Now, about those details . . .

What Mechanics Is All About

The study of the world around you requires knowledge of many areas of physics, often referred to as mechanics. The mathematician Archimedes of Syracuse (287–212 BC) is often credited as being the first person to systematically study the behavior of objects by using mechanics and is attributed with saying "Give me a place to stand and I will move the Earth." This statement, while rather grandiose for his time, proves itself to be at the very heart of the study of mechanics (and, more specifically, statics).

Mechanics refers to one of the core areas of physics, usually concentrated around the principles of Sir Isaac Newton and his basic laws of motion, and is an area of concentration that engineers and scientists often study in addition to basic physics classes. These courses develop the core curriculum for many basic engineering programs and are usually common classes across all disciplines. Specific engineering disciplines may require additional courses in each of these core areas to teach additional (and often more advanced) topics.

One of these core areas is in the area of *statics,* which isn't the study of how you should move across a shag carpet in order to apply a jolt of electricity to your younger siblings or how to implement the latest hygiene techniques to avoid those dreadful bad hair days. In this book, I define *statics* as the mechanical study of the behavior of physical objects that remain stationary under applied *loads* (which I discuss later in this chapter). The behavior of the floor beams in your house as you stand in the middle of your living room is an example of a static application.

The area of *dynamics*, on the other hand, is the study of objects in motion. So as you walk down the hall, your behavior and its effect on your house becomes a dynamic application. The result of a car driving down a bumpy road, the flow of water through a creek, and the motion of those shiny little metallic balls that hang from strings and haunt/hypnotize you with their "clack, clack, clack" as they bounce off each other are all examples of dynamic behavior.

Finally, you come to *mechanics of materials* (sometimes referred to as *strength of materials*), which is yet another branch of mechanics that focuses on the behavior of objects in response to loads. This area of mechanics builds on concepts from both dynamics and statics.

Putting Vectors to Work

One of the most basic tools to include in your basket of statics tricks is the knowledge of vectors, which I discuss in detail in Part II of this book. Think of vectors as being one of the major staples, such as rice or potatoes, of your statics kitchen. Statics forms the foundation for a complete meal of engineering design. Vectors come in all shapes and forms, and you can use them for a wide variety of purposes, which I introduce you to in Chapters 4 and 5.

But the vector discussion doesn't end there. I also show you several different ways to mathematically work with vectors, including building the foundation for a vector's equation (see Chapters 6, 7, and 8).

Peeking at a few vector types

One of the first vectors you need to get familiar with is the *position vector,* which basically tells you how to get from one point to another. These vectors are very handy for giving directions, measuring distances, and creating other vectors; you can read about them in Chapter 5.

The most common type of vector that you deal with has to do with loads, or forces (see the following section). Think of a force as being that number that pops up when you step on your bathroom scale that reminds you that you should have worked out last night instead of eating a second helping of cheesecake. The bigger that number gets, the bigger the force that is being applied to your scale. Forces are one of the major types of actions that can affect a body in statics.

Understanding some purposes of vectors

One purpose of vectors is to help define direction. Many forces act along straight lines but aren't necessarily acting at a distinct point. By creating a *unit vector* (a special type of vector with a specific length), you can define the direction of these lines without actually knowing the specific coordinates or location data; unit vectors also prove to be very useful for creating forces (another type of vector). Check out Chapter 5 for more on these vectors as well.

You can also use vectors to define the *rotational behaviors* (or spinning effects) of an object, which I explain in Chapter 12.

You can also combine multiple vectors to create a single combined vector, which can be useful for dealing with multiple forces. In addition, knowing how you can break down vectors into smaller vectors and calculate their size allows you to determine, say, how big a chair needs to be to support a given weight, including figuring out the size of its legs and even the number of legs necessary. In fact, for three-dimensional statics problems, vectors are practically mandatory. Chapters 7 and 8 deal with combining and breaking down vectors, respectively.

Defining Actions in Statics

In mechanics, you must become familiar with a large number of actions to be able to study how an object behaves, ranging from velocity and momentum in dynamics, to thermal effects, stress, and strain in mechanics. Fortunately, the types of effects in statics are contained in a fairly brief list:

- **Forces:** *Forces* are a type of load that causes an object to *translate* (move linearly) in the direction of the applied force. Forces can be spread out or acting at a single location, but they always cause an object to want to translate. You can use forces to measure the intensity of one object striking another, the weight of a car as it drives across a bridge deck, or the effect of water pressure on the side of a submarine. Flip to Chapters 9 and 10 for more on forces.

- **Moments:** *Moments* are a type of load that causes an object to rotate in space without translation. Moments are usually the result of some sort of twisting or spinning effect, such as a shaft attached to a motor, or a reaction from a second object that is acting on the other. For example, turning the handle of a wrench applies a moment to a bolt, which then causes it to rotate. Chapter 12 gives you the lowdown on moments.

One of your biggest challenges in statics is how to accurately depict and determine the type of action or behavior being applied to a system. If an elephant sits on your favorite living room recliner, you can easily tell what the final outcome of that action will probably be: You now have a broken chair, and a trip to the furniture store is in your future. Although most people will wonder how you got an elephant in your living room in the first place, as a statics enthusiast you're more interested in exploring the behavior of the elephant's weight and determining how much force is transmitted through the seat, into the legs, and ultimately into the ground. This field is where your study of statics begins (don't worry, no zoology or elephant anatomy knowledge is required).

Because forces and moments are such an important part of statics, you need to be able to calculate them for different kinds of problems. In Part III, I show you how to calculate forces and moments in both two- and three-dimensional situations. Load effects in statics are typically classified into three basic categories:

- **Concentrated forces:** *Concentrated forces* (or forces that act at a single point) include the force from a ball as it's thrown toward a wall, or even the force that your shoes exert on the floor from your self weight. I cover these forces more in Chapter 9.

- **Distributed forces:** *Distributed forces* are forces that are spread over an area and are used to represent a wide variety of forces on objects. The weight of snow on the roof of your house or of soil pressure on your basement wall is a distributed load. Chapter 10 shows you how to determine their net effect (or the *resultant*), and Chapter 11 illustrates how to determine the location where this resultant is acting.

✔ **Concentrated moments:** *Concentrated moments* are a type of load that causes a rotation effect on an object. The behavior of your hand on a door knob or a wrench on a nut is an example of rotational behaviors that are caused by moments. I describe the types of moments and how they are created in more detail in Chapter 12.

Sketching the World around You: Free-Body Diagrams

The ability to draw a *free-body diagram* (or F.B.D., the picture representations of the problem you want to investigate) is vital when you start a static analysis because F.B.D.s depict the problem you're trying to solve, and they help you write the equations you need for performing a static analysis. In fact, if you don't get the F.B.D. completely correct, you may end up solving for a completely different problem altogether.

The more you practice creating free-body diagrams, the more proficient you become. Free-body diagrams must feature various items, including dimensions, self weight, support reactions, and the various forces I discuss in Part III. (Head to Chapter 13 for a full checklist of required items.) You can also break a larger F.B.D. into additional diagrams; this tactic is useful because you can use these smaller diagrams to find information that helps you solve for a wide variety of effects, such as *support reactions* (physical restraints) and internal forces, that you may not notice on the larger drawing. You can find information on these topics in Chapter 14.

When you're working with F.B.D.s with multiple applied loads and supports, simplifying those diagrams can make your work a lot, well, simpler. Chapter 15 gives you several tricks for simplifying F.B.D.s; one of the most useful techniques is the *principle of superposition,* which allows you to quickly compute behaviors by simply adding the responses of the individual cases. You can also simplify your diagram by moving a force from one location on an object to another while preserving the original behavior; you can read more about this in Chapter 15 as well.

Unveiling the Concept of Equilibrium

Isaac Newton (1642–1727) helped establish the laws of motion and gravity (covered in Chapter 16) that are still used today. *Equilibrium* is a special case of Newton's laws where acceleration of an object is equal to zero (that is, it isn't experiencing an acceleration), which results in an object being in a stable or balanced condition. If you lean back in your chair such that it's supported by two legs, you notice that you reach a special point where you remain somewhat balanced. (But don't try this at home.) However, if you lean a little bit forward, the chair starts to rock forward and usually winds up safely back on the front two legs. This simple motion means that equilibrium hasn't been maintained. If you lean too far back, the chair starts to lean backward and unless you catch yourself, you soon find yourself lying on the ground. But good news: While you're lying on your back counting the little birds circling your head, you've actually arrived at a new equilibrium state.

Although you can simplify statics down to three basic equilibrium relationships for two-dimensional problems (and six equations for three-dimensional problems, though they're similar in concept), you can investigate a wide variety of problems with these relationships. Flip to Chapters 17 and 18 for more on equilibrium in two and three dimensions, respectively.

Applying Statics to the Real World

So what's an engineer to do after getting a handle on F.B.D.s, loads, equilibrium, and other statics trappings? Why, put them to use in actual applications, of course!

Real-world statics is where all the conceptual info you read about becomes much more interesting and much more practical. You can employ statics concepts to a wide variety of applications; some of the most common ones include the following:

- **Trusses:** *Trusses* are systems of simple objects interconnected to create a single combined system. They're commonly used in roof systems and as bridges that span large distances. In Chapter 19, I explain the basic assumptions of trusses and then illustrate the *method of joints* and the *method of sections* for analyzing forces within the truss.

- **Beams and bending members:** The majority of objects you work with in statics have up to three different types of internal forces (axial, shear, and moment, which I cover in Chapter 20). These internal forces are what engineers use to design structural members within a building. The forces sometimes cause a member to *deflect* (move away from being parallel), creating a *bending member.* You analyze these bending members by using shear and moment diagrams, which you can also read about in Chapter 20.

✔ **Frames and machines:** Frames and machines, though similar to trusses, can experience similar behaviors to beams and bending members. In fact, a large number of structural objects and tools that you use on a daily basis are actually either a frame or a machine. For example, simple hand tools such as clamps, pliers, and pulleys are examples of simple machines. *Frames* are more general systems of members that you can use in framing for structures. Chapter 21 gives you the lowdown on working with frames and machines.

✔ **Cable systems:** *Cable systems* are a unique type of structure and can produce some amazing architectural bridges known as *suspension bridges.* In Chapter 22, I describe the assumptions behind cable systems and present the techniques you need to solve cable problems.

✔ **Submerged surfaces:** *Submerged surfaces* are objects that are subjected to fluid pressure, such as dams. Fluids can apply hydrostatic pressure and pressure from self weight to submerged surfaces, and I describe both of those in Chapter 23.

A discussion of statics applications wouldn't be complete without talking about *friction,* the resistance an object feels along a contact surface as it moves in a particular direction. The two main types of frictional behavior are *sliding* (where the object moves across the surface in response to a force) and *tipping* (where the object responds to a force by toppling over rather than moving across a surface). These friction forces are the source of a large number of strange behaviors and require you to make assumptions about a behavior and then use free-body diagrams and the equations of equilibrium to verify them. Chapter 24 is your headquarters for all things friction.

Chapter 2

A Quick Mathematics Refresher

● ●

In This Chapter

▶ Paying attention to numerical accuracy

▶ Remembering numerical nomenclature

▶ Reviewing basic algebra

▶ Revisiting trigonometry and geometry identities

▶ Dealing with derivatives in calculus

● ●

*B*ecause you've chosen to study statics, you're probably already aware of the importance of mathematics in your studies. If not, I've got bad news: You just can't easily avoid numbers and calculations (particularly geometry and trigonometry) in your pursuit. Statics can provide you with solid physical principles for studying the world around you, but your skill with numbers and computations is what makes this information truly shine.

Think about the first rocket scientists that helped send astronauts to the moon with the Apollo missions in the 1960s and 1970s. Those scientists were among the smartest people on the planet at the time in physics theory, astronomy, dynamics, and statics (yes, even back then), among countless other areas of expertise. Imagine the success, or lack thereof, they would have experienced if they didn't have strong mathematical backgrounds.

This chapter reviews some of the mathematics skills required to efficiently solve statics problems. In this chapter I cover some basic nomenclature involving scientific notation, show algebra skills that prove useful, and review several geometric and trigonometric fundamentals. I conclude the chapter by showing how to use the power rule in calculus to integrate and differentiate.

Keeping Things Accurate and Determining What's Significant

In engineering, the accuracy of calculations can be the difference between a successful project and one that results in a pile of rubble on the ground. In all calculations, the numeric results have a certain number of digits that are meaningful, and some that serve no purpose other than to describe the magnitude of the number.

A *significant digit* is a nonzero value in any numeric quantity. *Nonsignificant* digits are those additional zeroes that help determine the magnitude of the number (such as the trailing zeroes on the number 2,500,000 or the leading zeroes on the decimal number 0.000156). The exception to this rule is the case of a zero digit that appears between two nonzero digits, such as in the number 106. In this case, the number 106 has three digits, all of which (including the zero) are considered significant.

Rounding is used to truncate irrational numbers. For example, the decimal form of π (pi) is 3.14159. . . . Because irrational numbers may have an infinite number of digits if you carry out the calculation far enough (which is the case of pi), you commonly round off the value to a specified number of places. For example, rounding 3.14159 . . . to two decimal places results in the value 3.14, which contains three significant digits. Similarly, rounding it to four decimal places gives the value 3.1416, or five significant digits.

Don't confuse significant digits with decimal places; the 3.14 estimation of pi contains three significant digits but only two decimal places.

One further complication in the accuracy of calculations involves figuring out how many significant digits you need in mathematical operations involving both extremely large numbers and very small numbers in the same calculation. Unfortunately, in many engineering calculations, you never know the number of significant digits you need to accurately represent a value until after you complete the calculation. However, if you keep a couple of basic rules of thumb in mind, you should be fine:

- ✔ **Multiplication and division:** When multiplying two numbers that have a different number of significant digits, remember that the final result should have as many significant digits as the number with the smallest number of significant digits in the original calculation.

- ✔ **Addition and subtraction:** When adding or subtracting two numbers that have different numbers of significant digits, remember that the final result should have as many significant digits as the number with the smallest number of significant digits in the original calculation.

Basically, no calculation result can have more significant digits than any of its original input values. For example, say you want to add these numbers:

123,456.789 + 0.000123456789 = 123,456.789123456789

The first value (123,456.789) contains 9 significant digits. The second value also contains 9 significant digits (remember, the preceding zeroes don't count). However, accurately reflecting the final sum of these two digits would require a staggering 18 significant digits to record precisely. But because 9 is the most significant digits in either term, that's the number you include in your answer.

In this book, I try to carry at least three decimal places in all cases regardless of the number of significant digits involved in the calculation.

Nomenclature with Little Superscripts: Using Scientific and Exponential Notation

A popular system of reporting numerical quantities for engineers and scientists is *scientific notation*. This method of representing numbers is very useful for briefly stating numbers that are extremely large or small. Scientific notation uses the base power of ten to greatly shorten written numbers by using a combination of a numerical multiplier and a 10 raised to some exponential value.

By employing scientific notation, you eliminate a lot of unnecessary scribbling. For instance, suppose you want to measure the distance to Pluto (either in planet or non-planet form) from the Earth in miles. Now before I start to receive nasty e-mails from my astronomer friends, I concede that this distance is actually highly dependent on where both planets are located on their current orbital cycle (among other factors). But, for the sake of this argument, say that this distance is roughly 2.7 billion miles.

In regular notation, you can represent this number as 2,700,000,000 miles. Clearly, this large number of zeroes is unwieldy. Using scientific notation, you can simply report this number as 2.7×10^9 miles.

The first term is the *numeric multiplier* and contains all the nonzero terms of your number. You always place the decimal just to right of the first nonzero digit (meaning the multiplier never goes higher than the ones place). After the multiplication sign, the next term is always the *10's multiplier*. To determine the *exponential power* (the little superscript number attached to the 10) on the multiplier, you need to count the number of decimal places required to move the decimal from its starting location in the original number to a position just to the right of the first nonzero numerical value. In this case, you move the decimal a total of nine spaces to the left (to land between the 2 and the 7), so the exponent is a 9.

For much smaller numbers, you move the decimal to the right, which results in a negative exponent. For example, Planck's Constant (which is a value used to describe the size of quanta in quantum mechanics) is expressed as 6.62606×10^{-34} N · m · s. This number written in normal notation has 33 preceding decimal zeroes before the actual numerical values. Add the extra place you shift to move the decimal to the right of the first nonzero numerical value, and you have a total exponent of –34.

The other basic rule you need to keep in mind is that when multiplying two values expressed in scientific notation, you multiply the numerical multipliers, add the exponent portions, and then make any final adjustments to the decimal placement to insure that you have only one nonzero value to the left of the decimal place. When multiplying exponent values with the same base (the 10 in this example), you simply add the exponents.

Suppose you want to multiply the distance to Pluto by Planck's constant. (Honestly, I'm not sure when you'd need that calculation, but it's always good to be prepared!) First, you multiply the numerical multipliers and then compute the new tens exponents by adding them, as follows.

$$(2.7 \text{ miles} \cdot 6.62606 \text{ N} \cdot \text{m} \cdot \text{s}) \times (10^{(9 + (-34))}) =$$
$$17.890362 \times 10^{-25} \text{ miles} \cdot \text{N} \cdot \text{m} \cdot \text{s} = 1.7890362 \times 10^{-24} \text{ miles} \cdot \text{N} \cdot \text{m} \cdot \text{s}$$

That comes out to $17.890362 \times 10^{-25}$ miles · N · m · s. But you can have only one nonzero number to the left of the decimal point, so you have to adjust your answer to $1.7890362 \times 10^{-24}$ miles · N · m · s. Of course, you should probably do some serious unit simplification on this answer — I cover that in Chapter 3.

Recalling Some Basic Algebra

In the world of statics, a few algebra skills can help you with some of the heavier lifting statics requires. In this section, I talk briefly about several common and useful algebra techniques you encounter in practice (and throughout this book). For more on these and other algebra topics, check out Mary Jane Sterling's *Algebra I For Dummies* and *Algebra II For Dummies* (Wiley).

Hitting the slopes of functions and lines

One of the more convenient mathematical tricks you use in statics is determining the slope and equation of a line between two points. The *slope* of a line is the ratio of the change in elevation (or height) to the change in horizontal distance; you may know it more simply as "rise over run." The equation for slope (signified by m) is $m = \dfrac{\Delta y}{\Delta x}$. You can also express this equation as $m = \dfrac{y_2 - y_1}{x_2 - x_1}$, where y_1, y_2, x_1, and x_2 are the coordinates of the two points.

For example, suppose that I rest a ladder on the ground at location Point 1 and on a ledge at location Point 2 (as shown in Figure 2-1) and want to define the properties of the line that connects these two points (the slope).

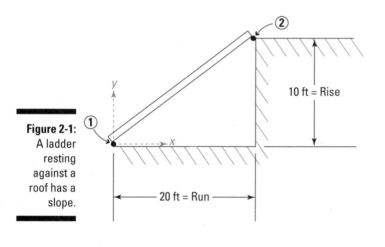

Figure 2-1:
A ladder resting against a roof has a slope.

$$\text{slope} = m = \frac{\Delta_y}{\Delta_x} = \frac{y_2 - y_1}{x_2 - x_1} = \frac{\text{rise}}{\text{run}} = \frac{10 - 0}{20 - 0} = \frac{1}{2} = 0.5$$

You may also remember that you can use slope to define the equation of the line passing through those two points, where b represents a constant for the y-axis intercept (or the point where the line crosses the y-axis at $x = 0$):

$$y = m \cdot x + b = \left(\frac{y_2 - y_1}{x_2 - x_1} \right) \cdot x + b$$

For the ladder example, you can solve for the numerical constant b by plugging in the values for either (x_1, y_1) or (x_2, y_2), both of which produce the same value for the constant. **Note:** The arrow in the following equation just indicates that I'm skipping some basic math.

$$10 = \left(\frac{10 - 0}{20 - 0} \right) \cdot (20) + b \Rightarrow b = 0$$

The constant b is dependent on where your measurement reference (or the origin — in this case, Point 1) is located. Thus, the equation of the line between Point 1 and Point 2 in Figure 2-1 is $y = 0.5x + 0$, simplified to $y = 0.5x$.

Calculating the slope proves to be a very handy trick when you deal with position vectors, which I cover in Chapter 5. After you have the equation of a line or function, you can make use of all sorts of cool mathematical tricks, which I point out in later chapters.

Rearranging equations to solve for unknown variables

Sometimes the equation you have doesn't solve for the variable you need. In that case, you can use algebra to juggle the equation in a way that suits it to your needs.

Suppose, for example, that you're given the following equation for the moment of inertia (I) of a rectangle having width b and height h.

$$I = \frac{1}{12}bh^3$$

This relationship would be handy if you knew both b and h. However, suppose you know h and I and want to find b. Rearranging this equation produces a different equation that would be more helpful in this scenario:

$$b = \frac{12I}{h^3}$$

Similarly, if you know b and I, you can rearrange the equation to solve for h:

$$h = \sqrt[3]{\frac{12I}{b}}$$

In this book, I provide guidance on how to generically solve a statics problem to produce a final equation. After you have the general equation, you can always rearrange the terms (following proper mathematical protocol, of course), to solve for a specific variable, or to create completely new relationships.

Sigma notation

Sigma notation (also sometimes called summation notation) is another popular form of shorthand notation; it utilizes the Greek symbol sigma (Σ), hence the name. Simply put, any time you see sigma notation, you know that you're about to do a whole lot of addition.

Suppose you have several variables (such as P_1, P_2, P_3, and P_4) that you want to add. You can simply write that equation as follows:

$$P_{TOTAL} = P_1 + P_2 + P_3 + P_4$$

That isn't too bad an expression, so long as you don't want to add an extremely large number of terms. But what if you had a list of a thousand or a million terms? You'd need several sheets of paper and a whole lot of time to complete that problem! Imagine writing that expression as

$$P_{TOTAL} = P_1 + P_2 + P_3 + \ldots + P_{n-1} + P_n$$

where n is equal to however many terms you have. Sigma notation allows you to conveniently express this type of equation in a single compact method:

$$P_{TOTAL} = \sum_{i=1}^{n} P_i = P_1 + P_2 + P_3 + \ldots + P_{n-1} + P_n$$

This is sigma notation. The sigma indicates that this expression is an equation involving addition. The variable below the sigma represents a counter and is increased by one from the first term (in this example, 1) each time you add a term to the expression. The variable above the sigma indicates the value of the final term in the expression (or n in this example). If you wanted to add only the terms from P_3 through P_n, you'd just change $i = 1$ to $i = 3$.

If you wanted to rewrite the first example of P_1 through P_4, you need only modify the variables above and below the sigma, again:

$$P_{TOTAL} = \sum_{i=1}^{4} P_i = P_1 + P_2 + P_3 + P_4$$

Pretty simple and compact, huh? In fact, in statics, the fundamental equations of equilibrium utilize sigma notation every time you write an equation. I explain more about equilibrium in Part V, but a word to the wise: Get familiar with sigma notation now, if you haven't already!

Getting into Shapes with Basic Geometry and Trigonometry

Algebra isn't the only basic math you encounter in statics (see the preceding section). You also use some basic geometry and trigonometry principles on a regular basis, so the following sections give you the lowdown on these concepts.

Getting a handle on important geometry concepts

In this section, I introduce a couple of geometric relationships that show up frequently in statics. They show you how you can compute the total angles contained within a polygon as well as how to relate angles created by the crossing of multiple lines. *Geometry For Dummies*, 2nd Edition, by Mark Ryan (Wiley) gives you more detail on geometric concepts.

Computing angles inside polygons

In any physical analysis problem, basic geometry often plays a very important role. Statics is no different — in many of the analysis problems you encounter, you need to make use of several basic geometric relationships.

The first relationship involves determining the total degrees in a polygon of a given number of sides. Triangles are very popular shapes within static analysis, as are quadrilaterals, parallelograms, and other higher-order shapes.

The sum of the interior angles for a polygon having n sides can be given by the expression

Total Degrees in Polygon = $180(n - 2)$

You can easily confirm this formula by using your basic knowledge of triangles and quadrilaterals. A triangle has three sides ($n = 3$) and a total of 180 degrees: $180(3 - 2) = 180(1) = 180$. Similarly, a quadrilateral has four sides ($n = 4$) and a total of 360 degrees. Quadrilaterals and triangles (such as those in Figure 2-2) account for the majority of the problems in this book, but on occasion you need to venture to more-complex shapes, and this rule proves handy for those cases.

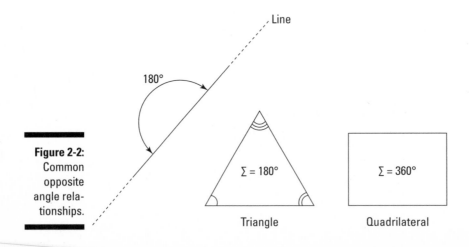

Figure 2-2: Common opposite angle relationships.

Line

180°

$\Sigma = 180°$

Triangle

$\Sigma = 360°$

Quadrilateral

Constructing angles created from line segments

Another relationship involves geometric constructions with two parallel lines, *A* and *B,* and a third line *C* that crosses them both (see Figure 2-3). In this figure, θ_3 and θ_5 are *opposite interior angles* (or angles that are across from each other whenever two lines cross), and θ_4 and θ_6 are also opposite interior angles.

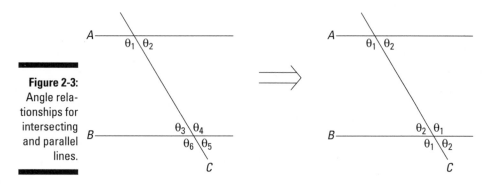

Figure 2-3: Angle relationships for intersecting and parallel lines.

In this construction, you can see that the angles θ_1 and θ_2 comprise the complete 180 degrees of line *A.* That is,

$$\theta_1 + \theta_2 = 180 \text{ degrees}$$

Similarly, along line *B,*

$$\theta_3 + \theta_4 = 180 \text{ degrees, and } \theta_5 + \theta_6 = 180 \text{ degrees}$$

Along line *C,*

$$\theta_1 + \theta_3 = 180 \text{ degrees, } \theta_4 + \theta_5 = 180 \text{ degrees, and } \theta_3 + \theta_6 = 180 \text{ degrees}$$

These three constructions thus imply that

$$\theta_1 = \theta_4 = \theta_6 \text{ and}$$
$$\theta_2 = \theta_3 = \theta_5$$

Double-checking angles with degrees and radians

One of the more common hang-ups I see when working with angular measurements is the basic confusion that exists between the units *degrees* and *radians.* It turns out that both of these base units are actually related to each other, as the following formula shows. Recall that a circular shape has 360 total internal degrees. In radians, this same internal angle is represented by 2π radians. (Remember that $\pi = 3.14159.\ldots$)

$$1 \text{ radian} = \frac{360°}{2\pi} = 57.296\ldots°$$

You definitely want to pay special attention to which unit setting your calculator is currently working with — most calculators are capable of dealing with both degrees and radians, and many calculators can be easily (and accidentally) switched between these two modes. In fact, I've experienced many a calculation going awry because I failed to switch modes. But if you're careful, this mix-up won't be a major issue. Be sure to consult with your calculator's instruction manual if you're having issues with switching between units.

Recalling the Pythagorean theorem

The Pythagorean theorem is another useful geometric relationship that allows you to relate the sides of a *right triangle* (a triangle with one angle of exactly 90 degrees). Consider the right triangle shown in Figure 2-4.

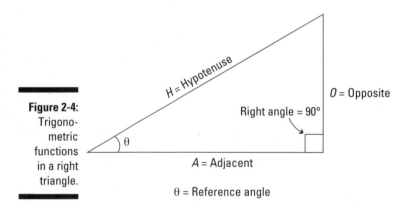

Figure 2-4:
Trigono-
metric
functions
in a right
triangle.

H = Hypotenuse

O = Opposite

Right angle = 90°

θ

A = Adjacent

θ = Reference angle

You may have seen the formula written as $C^2 = A^2 + B^2$ or $H^2 = A^2 + O^2$. Regardless of which letters you use, this formula relates the *hypotenuse* (the side opposite of the 90 degree angle) to the two other sides. This formula is very useful when you when you start working with vector resultants (see Chapter 7).

Tackling the three basic identities of trigonometry

Trigonometry is the branch of mathematics that deals with triangles. The cornerstones of trigonometry are the sine, cosine, and tangent functions that define the relationships among the sides of a right triangle. Referring to Figure 2-4, you can see that Side *A* is the side adjacent to the reference angle θ, and Side *O* is the side opposite to the reference angle. Finally, Side *H* represents the hypotenuse of the right triangle and is found directly across from the right angle.

To help you remember these relationships, try using the anagram SOHCAHTOA. No, Sohcahtoa wasn't the guide who helped Lewis and Clark explore the western frontier and ultimately discover the Pacific Northwest. SOHCAHTOA can, however, be a tremendous guide for remembering the three basic identities of trigonometry:

$$\text{SOH} \Rightarrow \boxed{\text{s}}\text{in}(\theta) = \boxed{\frac{O}{H}} = \frac{\text{opposite}}{\text{hypotenuse}}$$

$$\text{CAH} \Rightarrow \boxed{\text{c}}\text{os}(\theta) = \boxed{\frac{A}{H}} = \frac{\text{adjacent}}{\text{hypotenuse}}$$

$$\text{TOA} \Rightarrow \boxed{\text{t}}\text{an}(\theta) = \boxed{\frac{O}{A}} = \frac{\text{opposite}}{\text{adjacent}}$$

The hardest part is just remembering how to spell it! S-O-H-C-A-H-T-O-A!

You want to be sure to carefully denote which angle of the right triangle is your reference angle, because its location can affect your assignment of *O* and *A* in those expressions.

Brushing Up on Basic Calculus

A few of the basic calculus skills that may come in handy in your statics work include the differentiation and integration of polynomials and the locations and value of maximum and minimum values of polynomial functions. Luckily for you, I discuss both in the following sections. Check out Mark Ryan's *Calculus For Dummies* and Mark Zegarelli's *Calculus II For Dummies* (Wiley) for a complete calculus review.

The power rule: Differentiation and integration of polynomials

Before I illustrate a few of the simpler basics of calculus, keep in mind that there is significantly more to differentiation in calculus than just the power rule. After all, most engineers and scientists are required to take multiple semesters (sometimes three or four) of various levels of calculus to complete their degrees.

That being said, a large portion of the content covered in a basic statics course can be encompassed with the power rule, so that's where I start.

Basic differentiation and tangents to functions

The *derivative* of a function represents the slope of the tangent line to the function at a particular location. The derivative of a constant is always zero. For a simple function $f(x)$, I define the derivative as $f'(x)$, which is equivalent to $df(x)/dx$. (In this case, the derivative represents the slope of the tangent line to the function at x.) The *power rule* states that for a smooth and continuous polynomial of order n, the derivative of a function $f(x)$ can be expressed as

$$f(x) = x^n \text{ then } f'(x) = \frac{df(x)}{dx} = n \cdot x^{n-1}$$

The order of a polynomial determines the shape of the curve. A zero order polynomial is constant, a first order polynomial is linear, and a second order polynomial is curved (or more specifically quadratic).

For example, for the function

$$f(x) = 4x^3 + 5x^2 + 24$$

you can compute the derivative of as

$$f'(x) = 4 \cdot (3) \cdot x^{(3-1)} + 5 \cdot (2) \cdot x^{(2-1)} + (0) \cdot 24 = 12x^2 + 10x$$

The terms inside the parentheses indicate the powers of the original term being differentiated. Because the derivative of a constant is always zero, the 24 in this equation disappeared.

The examples I show here are for first derivatives, but you can also have higher-order derivatives, such as second, third, or even hundredth derivatives, in calculus. To compute the second derivative of a function, you compute the first derivative as I explain here and then compute the derivative of that derivative. The higher the order derivative that you want to compute, the more derivatives you have to take. Fortunately, in statics, usually a second or third order derivative is sufficient.

Basic integration

For basic integration, a definite integral for a simple polynomial can employ a reverse process to the differentiation technique for the power rule. The following equation assumes that the polynomial function $f(x)$ is smooth and continuous and evaluated between an upper limit b and a lower limit a.

$$f(x) = \int_a^b f'(x) \cdot dx = f(b) - f(a)$$

When you perform this calculation, you're actually finding the area under the function between the limits of a and b. This value can come in really handy when you start calculating centroids (see Chapter 11). To integrate a smooth and continuous polynomial of order n such that $f'(x) = x^n$, the integral becomes

$$f(x) = \int_a^b x^n \cdot dx = \frac{x^{n+1}}{n+1}\bigg|_a^b = \left(\frac{1}{n+1}\right)\left(b^{n+1} - a^{n+1}\right)$$

Using calculus to define local maximum and minimum values

On many occasions, the statics equations you write contain variables that are frequently in the form of *smooth and continuous polynomials* (meaning that the graph of the function doesn't contain any jumps or sharp changes) of some order n.

This setup is pretty convenient because the power rule I discuss in the previous sections works effectively on polynomials. The ability to be able to determine the locations of maximum and minimum values of a polynomial function is even handier. If you recall that the slope of a line tangent to a maximum or minimum value is always horizontal (or equal to zero), you shouldn't be surprised that

$$\frac{dy}{dx} = f'(x) = 0$$

In order to find the location of a local maximum or minimum value, all you need is the first derivative of the original function, the ability to set that first derivative equal to zero, and the ability to find the value(s) of the independent variable x that satisfy that equation. After you determine the locations, simply plug those x values back into the original function $f(x)$ and compute the value of that function. For example, consider a third-order function:

$$f(x) = x^3 + 5x^2 - 8x - 12$$

Setting the first derivative equal to zero allows you to find the locations x of the local maximum and minimum values.

$$f'(x) = 3x^2 + 10x - 8 = 0$$

For this equation, you can find that $x_1 = 0.667$ and $x_2 = -4.000$. Substituting these locations into the original function, you can determine which is the location of the local maximum value.

$f(0.667) = -14.815$ and $f(-4.000) = 36.000$

From this result, you can conclude that the local maximum value of the polynomial $f(x)$ is +36.000 and occurs at a location of $x = -4.000$.

Chapter 3

Working with Unit Systems and Constants

A s you drive down a road in the United States, you typically see the speed limit displayed in terms of mph, or miles per hour. However, this abbreviation has little or no meaning in many countries; instead, citizens of those countries may refer to speed limits in kph, or kilometers per hour. Both are perfectly acceptable units of measure, but because of local customs and preferences, familiarity with the other system of units may be somewhat lacking (so make sure you're checking the correct dial on your speedometer before you pass a speed trap!).

When working any statics problem, you see that the basic equations and relationships are consistent regardless of the measurement system. However, when you actually put those equations to work, the measurement systems and units play a very important role.

In this chapter, I unravel the two major systems of units and explain each in detail, including proper base units and metric system prefixes. I also provide you with some tips on how to convert between those two systems and conclude by discussing issues to remember regarding numeric computational accuracy. Though this chapter by itself can't get you ready to work a complete problem, it can hopefully give you a better awareness of the unit systems that run throughout statics.

Measuring Up in Statics

Before you start crunching all those wonderful numbers and creating all the awesome equations statics requires, you first need to be clear on what system of units you need. Depending on where you live, or whom you're working with, you need to understand each of the two basic sets of units that are commonly used: the metric system and U.S. customary units. Good news: I cover them both in the following sections.

The metric system

The *metric system* is a system of units that utilizes a base unit for everything from mass to forces to distances. Table 3-1 shows some of the more common base units and abbreviations you may encounter when using the metric system.

Table 3-1	Metric Base Units and Abbreviations	
Measurement	*Metric Units*	*Metric Abbreviations*
Length	meter	m
Force	Newton	N
Time	second	s
Mass	gram	g

I should point out a slight exception to the units of force for the Newton (N). The Newton is actually a *derived unit* created from a combination of other units and is expressed as

$$1 \text{ N} = \left(1 \text{ kg} \right) \cdot \left(1 \frac{m}{s^2} \right)$$

Notice how this expression contains a mass unit of kilogram, when the metric base unit for mass is actually in grams. The second term is a combined unit for acceleration. Remember, when you compute a force in Newton units, the mass needs to be expressed in kilograms. I show you how to convert grams to kilograms in "Converting to larger and smaller metric units" later in the chapter.

When working with metric units, you have to be able to convert between base units with different prefixes. In the later section "Converting to larger and smaller metric units," I show you how easy these conversions are.

The International System of Units is a system of standardized units that uses measurements exclusively from the metric system. The SI abbreviation is short for the French system *Système International d'Unités* and is used extensively in many parts of the world. Within the SI/metric system, you always need to be familiar with a subset of conversions.

Converting to larger and smaller metric units

After you select a proper base unit (see Table 3-1 earlier in the chapter), you attach a series of prefix values to that base unit to create a *scaled unit* (a unit that is either larger or smaller than the base metric unit). Table 3-2 shows some of the more common prefixes, including prefixes for getting larger (giga-, mega-, and kilo-) as well as prefixes for getting smaller (centi-, milli-, micro-, and nano-).

Table 3-2		Metric Conversions	
Prefix	**Symbol**	**Multiplier**	**Exponential Conversion**
Getting Bigger			
giga-	G-	1,000,000,000	10^9
mega-	M-	1,000,000	10^6
kilo-	k-	1,000	10^3
Getting Smaller			
centi-	c-	0.01	10^{-2}
milli-	m-	0.001	10^{-3}
micro-	μ-	0.000001	10^{-6}
nano-	n-	0.000000001	10^{-9}

To increase from a smaller prefix to a larger prefix, you must multiply by the exponential conversion shown in Table 3-2. The first term in the conversion is always the starting units. The second term is always the conversion to go from the starting units back to the base units. For example, to convert one kilometer to its base units of meters, you set up an equation like the following:

$$1 \text{ km} \cdot \left(\frac{10^3 \text{ m}}{1 \text{ km}} \right) = 1,000 \text{ m} = 1.0 \times 10^3 \text{ m}$$

    (1)   (2)

The first term (1) is the starting units, or kilometer in this example. The second term (2) is the starting-unit-to-base-unit conversion; the units in the numerator (or top) of the first term should always be the same as the units in the denominator (or bottom) of the second term. (***Note:*** This example's first term doesn't have a visual numerator, but 1 km is the same as $\frac{1 \text{ km}}{1}$, so 1 km acts like a numerator in this case.)

Making multiple conversions in one equation

After you master basic conversion (see the preceding section), you can also do multiple conversions all in the same step. The basic conversion formula looks like the following:

Starting Units · Conversion to Base Unit · Conversion to Final Unit = Final Units

All you have to do is simply chain together multiple conversion calculations like those in the previous section. For example, suppose you want to calculate how many milligrams are in one megagram. Your calculation would look something like this:

$$1\,\text{Mg} \cdot \left(\frac{10^9\,\text{g}}{1\,\text{Mg}} \right) \cdot \left(\frac{1\,\text{mg}}{10^{-3}\,\text{g}} \right) = 1.0 \times 10^{12}\,\text{mg}$$

   (1)   (2)   (3)

In this example, the starting units are megagrams, which means that base units are grams (as given by Table 3-1 earlier in the chapter). Your first order of business, then, is to convert from megagrams to grams by multiplying by 10^9 as shown in parts (1) and (2) of the equation. But you're not done yet — your final units are milligrams, so you need part (3) of the equation, which requires dividing the converted base value by 10^{-3} (because 1 milligram contains 10^{-3} grams). Remember to make sure the units in the numerator of (2) are the same as the units in the denominator of (3).

Watching the units and prefixes in this manner can greatly simplify your work. In fact, as long as the units end up in the proper denominator and numerator, you actually have a bit of flexibility in the conversion that you perform. For example, another way of looking at the final conversion of the previous example is by remembering that there are also 1,000 milligrams in 1 gram. With this information, you can rewrite the third part of that equation as

$$1\,\text{Mg} \cdot \left(\frac{10^9\,\text{g}}{1\,\text{Mg}} \right) \cdot \left(\frac{1,000\,\text{mg}}{1\,\text{g}} \right) = 1.0 \times 10^{12}\,\text{mg}$$

   (1)   (2)   (3)

Both calculations achieve the same desired result. Remember, when working with prefixes, let the units do the work for you!

U.S. customary units

The *U.S. customary units*, often referred to as *English units,* are the unit system used predominately in the United States. Like their metric counterpart, U.S. customary units also have commonly used base units, which I list in Table 3-3.

Table 3-3	U.S. Customary Base Units and Abbreviations	
Measurement	*U.S. Customary Units*	*U.S. Abbreviation*
Length	foot	ft
Force	pound	lb (or #)
Time	second	s
Mass	slug (1 lb sec^2/ft)	slug

The kip: One crazy exception

Typically, the metric and U.S. customary systems are completely exclusive. However, one major exception is the *kip,* a hybrid unit used to express very large forces and pressures. The kip unit is actually an abbreviation for the *kilo-pound,* which is a combination of the metric prefix kilo- and the U.S. customary unit of force, pounds. Because kilo- means 1,000, you can deduce that 1 kip equals 1,000 pounds. Though the kip isn't a true unit of measure in either system, it's very convenient because it allows you to write bigger numbers with fewer zeroes.

Some textbooks further abbreviate the kip to a single *k.* To keep from confusing this abbreviation with the metric prefix for kilo-, remember that kilo- is a prefix and is always followed by some other base unit of measure, such as gram, and the abbreviation reflects that unit (kg). On the other hand, kip is already a unit and therefore doesn't get anything else attached to its end — it typically follows a numerical measure (20 k, –134.27 k).

Never the twain shall meet: Avoiding mixing unit systems

When a situation arises where you have items expressed in different systems, your best bet is to go ahead and convert everything to the same system of units. This section shows you how to do just that.

When you find that you need to convert U.S. customary units to metric units (or vice versa), having some conversion factors can come in handy. With the help of Table 3-4 and the conversion process I outline earlier in the chapter, you can convert from system to system with no problem.

Houston, we have a unit problem

When working with units, you should always try to avoid combining units from the metric and U.S. customary systems because this move can often cause serious problems. It's what led to a failure of NASA's Mars Climate Orbiter in 1999. This instrument burned up on entry into the Martian atmosphere as a result of a computer conversion error in the calculation of thrust. In simple terms, the computer failed to properly convert between a force expressed in pounds and a force expressed in Newton.

Table 3-4	U.S. Customary to Metric Conversion Factors	
Measurement	*U.S. Customary Units*	*Metric Units*
Length	1 ft	0.3048 m
Force	1 lb	4.448 N
Mass	1 slug	14.59 kg
Acceleration	1 ft / sec^2	0.3048 m / sec^2
Energy	1 ft-lb	1.356 N − m = 1.356 J
Pressure and Stress	1 lb / ft^2	47.88 Pa

For example, say your problem deals with a distance expressed in feet (U.S. customary) and a force expressed in Newton (metric). You have two options:

✔ **Convert to metric:** To convert a U.S. customary foot to a metric distance, you perform the following conversion:

$$1 \text{ N} \cdot \text{ft} \cdot \left(\frac{0.3048 \text{ m}}{1 \text{ ft}} \right) = 0.3048 \text{ N} \cdot \text{m}$$

✔ **Convert to U.S. customary:** To convert a metric Newton to a U.S. customary measurement, you use this conversion:

$$1 \text{ N} \cdot \text{ft} \cdot \left(\frac{1 \text{ lb}}{4.448 \text{ N}} \right) = 0.2248 \text{ lb} \cdot \text{ft}$$

Looking at Units of Measure and Constants Used in Statics

The study of statics utilizes many of the units I discuss throughout this chapter. However, many of the units you encounter are combinations of the base units listed in Table 3-1 earlier in the chapter. This section discusses several major categories of those units and their appropriate metric and U.S. customary equivalents.

Constants worth noting

You use several physical constants on a regular basis in statics; I list a few of the most widespread in Table 3-5, and they appear throughout the book.

Table 3-5	Common Constants	
Measurement	*Metric Units*	*U.S. Customary Units*
Gravitational acceleration	9.81 m/sec^2	32.2 ft/sec^2
Dimensions	No conversion	1 ft or 12 inches
Specific weight of water	9810 N/m^3	62.4 lb/ft^3

Three common statics units for everyday life

Although units are always important in any calculation you perform, several categories of units seem to occur more often than others. Following are three of the more common base units you may encounter. As with all metric units, you may see the base units I give here with a different prefix, such as those in Table 3-2 earlier in the chapter.

- ✔ **Distance:** This category includes units that measure the length or dimensions of or between objects. In metric units, the standard unit is the meter; in the U.S. customary system, it's the inch or foot.

- ✔ **Angles:** An *angle* is a measurement of an orientation of one line segment with respect to another. The common units for angles are *radians* (which are a derived unit of 1 meter per meter or 1 foot per foot) and *degrees,* both of which are consistent in both the metric and U.S. customary systems.

✔ **Force:** A *force* is a type of action between physical bodies, or between a body and its environment. (I discuss forces in more detail beginning in Chapter 9.) Its standard units are the Newton (for metric) and the pound (for U.S. customary).

All the derived units you'll ever need

Several common statics units are based on calculations involving the base units listed in the preceding section. A few of the more commonly used units are as follows:

✔ **Moments:** A *moment* is an action that causes rotation (which I talk about more in Chapter 12). In metric units, the standard base unit for a moment is the Newton-meter (N-m), and in the U.S. customary system, the base unit is the foot-pound (ft-lb or lb-ft — the order doesn't matter).

✔ **Distributed force effects:** I cover the effects of forces acting over a given length more in Chapter 10; these units are expressed as a "force per distance." Here, you just need to know that their metric unit is Newton per meter (N/m) and their U.S. customary unit is pounds per foot (lbs/ft). Another common representation for lbs/ft is *plf,* which is an abbreviation for "pounds per linear foot." Similarly, in the event of larger forces, you may also encounter a unit of *klf,* or "kips per linear foot." Check out "The kip: One crazy exception" earlier in the chapter for more on the hybrid unit kip.

✔ **Pressure effects:** A *pressure* is a force that acts over a discrete (distinct) area. The metric unit for pressure effects is Newton per square meter (N/m^2). This unit is also known as the pascal and may be abbreviated as Pa. The U.S. customary representation is usually either pounds per square foot (lbs/ft^2 or psf) or pounds per square inch (lbs/in^2 or psi).

✔ **Volumetric effects:** A *volumetric effect* is a force (such as specific weight) acting on a given volume. In the metric system, their unit is Newton per cubic meter (N/m^3). The U.S. customary units are pounds per cubic foot (lbs/ft^3 or pcf).

Part II
Your Statics Foundation: Vector Basics

The 5th Wave   By Rich Tennant

In this part . . .

*V*ectors are a huge part of statics, so in this part, I explore the basics of vector mechanics by showing you how to depict a vector and explaining a vector's basic properties. I also show you how to actually create a vector based on position data, and then I demonstrate how you can use this information to create additional vectors. I illustrate how you can combine multiple vectors into a single resultant vector, as well as break a single vector into smaller pieces. As if all that weren't enough, I also give you the lowdown on basic vector mathematics.

Chapter 4

Viewing the World through Vectors

. . . .

In This Chapter

▶ Understanding basic vector terminology and properties

▶ Sorting through the types of vectors

▶ Depicting a vector

. . . .

*B*efore you can build a bridge or even think about completing a design, you have to begin by understanding the way engineers depict the world around them, a fundamental concept in statics. Enter the vector. Essentially, the study of vectors is the first step into this larger world of statics analysis.

This chapter focuses on exploring the behavior of vectors, seeing the commonalities in their construction, and understanding some of the subtle differences in their creation and application. In this chapter, I define the three major pieces of information you use to help a vector properly describe an action, I show you a few ways to draw a vector, and I break down the three primary types of vectors. This chapter won't have you building a bridge immediately, but it does help you take your first step in getting a handle on the world around you in proper statics style.

Defining a Vector

You quickly discover that the ability to create and define a proper vector is an invaluable set of skills. This ability lends itself fluidly to solving statics problems. However, before you can become truly proficient in statics, you first need to understand some basic terminology and the three pieces of information you need to properly define a vector.

Simply put, a *vector* is a quantity that helps describe the way that an action is applied to an object or group of objects. For example, a velocity vector can describe the velocity motion of a golf ball after it has been hit by a nine-iron, and a distance vector can help depict how far away and in what direction it landed. A force vector can describe how hard and in what direction the golf club strikes the golf ball.

Many different types of vectors exist — from velocity and displacement vectors, to vectors that describe magnetic field behavior, to vectors that are mathematical solutions of differential equations. In statics, the force vector is the main type of vector you have to deal with. (*Note:* Don't confuse these *types* of vectors with the *categories* of vectors I describe later in the chapter.)

Understanding the difference between scalars and vectors

Before you can dive very far into the heart of your study in statics, you need to understand the difference between a vector and a scalar, two terms that are always popping up in statics textbooks and practice:

- ✔ **Scalar:** A *scalar quantity* (or simply a *scalar*) is any measurement made only with regard to an action's amount (its *magnitude*) and not its direction. Examples of scalar quantities include the cost of this book, the temperature of the room around you, or the airspeed of Monty Python's unladen swallow. You can describe all these quantities as a single amount. Even time is considered a scalar quantity because time only moves in one direction (supposedly).

- ✔ **Vector:** As I mention earlier in the chapter, a vector is a quantity that describes both the size (an amount) *and* direction of a particular action. Examples of a vector include the approach flight path of an airplane coming in for a landing (the distance from the runway is a scalar quantity, and the flight path trajectory defines the direction), the velocity of a speeding car (the speedometer reading is a scalar, and the compass on the dashboard indicates the direction), or the force of an elephant sitting on a chair (the mass of the elephant is a scalar quantity, and the direction of gravity defines the direction of the elephant's force).

  In fact, a scalar quantity is often part of the information contained within a vector definition, but I talk more about that in Chapter 5. For example, the speed of a moving elephant (a scalar quantity) is directly related to the velocity vector of that mammal.

Taking a closer look at vectors

Although you can display a scalar entity simply by jotting a number or measurement on your piece of paper, the proper representation of a vector requires three pieces of very specific information:

✔ **Magnitude:** The magnitude is the numerical value of a given vector. Constructing a vector requires actually knowing a scalar quantity — the magnitude. Magnitudes of vectors are scalar quantities and may be positive, negative, or zero in value. Just remember that there is no direction associated with a magnitude. (Sit tight — I talk about how to actually create a vector from a scalar quantity in Chapter 5 and how to calculate the magnitude of a vector in Chapter 8.)

✔ **Sense:** The *sense* of a vector is the *sign of the magnitude,* or the direction in which the vector is acting. The sense is the part of a vector that indicates whether a charging elephant, moving at a speed of 20 miles per hour, is heading toward or away from you. You can describe that direction in several ways, and I cover each of them in Chapter 5.

✔ **Point of application and lines of action:** The *point of application* is the physical location on the object or in space where the vector is acting. The *line of action* of a vector refers to the line in space on which the vector is acting, regardless of whether the vector is acting internally or externally to the object. In all cases, however, the line of action of a vector and the vector's point of application (if it has one — certain types of vectors don't!) always coincide. That is, the line of action of a vector passes through the point of application.

Vectors can act either internally or externally:

✔ **External vector:** A vector that acts on the external surface of an object. Examples of external vectors include drag forces on the wing of an airplane, the friction forces you feel when you rub the palms of your hands together, and the force of your hands on the cover of this book as you read it. (I discuss external vectors in more detail beginning in Chapter 7.)

✔ **Internal vector:** A vector that acts on the object at a specific internal location. Examples of internal vectors include the weight of this book and the internal compressive force in the legs of the chair you're sitting in (if you're sitting). I dive deeply into the subject of internal vectors beginning in Chapter 16.

Knowing whether vectors are acting externally or internally can help you decide how to best tackle a given problem.

Applying vector basics

Sometimes, there's no substitute for a good example. In this section, I outline a scenario that I hope helps you get a firm grasp on vectors and their components (which I cover in the preceding section). Imagine trying to give driving directions to a friend travelling from his house to yours. In your discussion with him, you'd never have a list of directions that states

> "Okay, Tom, first go a half of a mile, and then go another three-quarters of a mile, and finally head a quarter of a mile."

The representation of this directional data is an example of a scalar quantity. All three distances in the previous statement are nonnegative values and are actually all considered magnitudes.

Unless you simply don't want visitors or you like getting frantic cellphone calls from lost friends (and admit it, who doesn't?), you need to provide significant information that's missing from that first list of directions.

In this case, you definitely need to describe the sense, or the direction, of each of those measurements. You can establish the sense of these directions by using relative descriptions such as "turn left" or "veer right," but this approach can be dangerous. If your intended direction is west and you tell someone accidentally heading south to turn left, his final direction will be completely opposite of where you want him to go. One wrong turn can render relative descriptions completely inaccurate. To avoid this dilemma, use absolute sense descriptions by giving cardinal directions with the proper instructions. As you begin to construct vectors, take special care to formulate your vectors with specific absolute information.

In Tom's case, a better set of the previous driving directions, incorporating absolute sense description, may be

> "Okay, Tom, first go a half of a mile north, and then go another three-quarters of a mile east, and finally head a quarter of a mile south."

However, even this set of instructions is still lacking a significant piece of specific information. In this case, though the directions themselves are decent and definitely more detailed, you still don't know where the trip starts. Tom's starting point in this example represents the point of application.

The current instructions are adequate for relative positioning, although the final destination becomes directly dependent on the starting point — if you change the starting point, the final destination obviously changes as a result. You can vastly improve these relative directions if you also mention the starting position:

> "Okay, Tom, from your home, first go a half of a mile north, and then go another three-quarters of a mile east, and finally head a quarter of a mile south. This will get you to my house."

From this list of directions, your visitor should have very little trouble moving from his house to yours (as long as he doesn't hit any construction zones or detours along the way)!

Drawing a Vector's Portrait

Unfortunately, in describing driving directions for how your friend travels from his house to yours, you don't usually see the path as a line on the ground or on the map (unless of course his car has an oil leak). Similarly, you can't physically see a force vector in action; you see only its resulting influence on the object. Because engineers are always making sketches to help describe the world around them, you need to be familiar with the techniques they use to graphically depict a vector.

To draw a vector, you have to graphically represent the three major components of a vector: magnitude, sense, and point of application, which I discuss in the preceding section. In the sections that follow, I explain how you graphically represent this information when drawing a vector, as well as describe two common vector depictions.

The single-headed arrow approach

In this book, as in most statics and mathematics references, I typically represent vectors as single-headed arrows that break down into a number of parts (which you can see in Figure 4-1):

- **Head:** The arrowhead indicates the vector's sense. However, an exception to this guideline can occur when vectors are pushing on an object; in this case, the head of the arrow is commonly used to indicate the point of application. I explain this exception in more detail in Chapter 9.

- **Tail:** The tail of the arrow typically depicts the vector's point of application (barring the exception described in the preceding bullet).

- **Shaft:** The actual line-length of the arrow represents the vector's magnitude — a longer vector drawing implies a larger action and vice versa. The shaft of the arrow aligns with the axis or line of action of the vector. Vectors that aren't oriented horizontally or vertically sometimes include an angle measurement to help define the vector's orientation. This angle is usually measured from either a horizontal or vertical reference. I discuss vector notation more in Chapter 5.

- **Label:** The label of the vector can be a letter or name given to a particular vector arrow to help distinguish it from other vectors in your drawings. Sometimes, you actually write the value (with proper units, of course) of the magnitude of the vector as the label. In other situations, you may use an alphanumeric label such as Load1, WindForce, or Bob'sWeight — the sky is the limit on how you actually label your vectors, but it helps if you name it something to remind you of what that arrow represents on the picture. To help distinguish these entities, some textbooks (and this book) write nonnumeric vector labels in bold: **V**. Other texts commonly display vector labels with an arrow symbol over the top: $\overline{V}$.

This label serves two basic purposes:

- It acts as a reminder of specific scalar information that you may have calculated previously or already know from a statement given in the problem. If you don't know this specific piece of information at the time (and you often don't), you can use the name or label of this vector as variables in your equations.

- For the sake of convenience, you often don't include the magnitude of the vector graphically (by making a vector longer or shorter) in your depiction. Instead, you write the numerical quantity of the magnitude (if it's known) beside the arrow. Doing so saves space when you're constructing free-body diagrams (which I dive into in Part IV) and helps really small vectors remain visible. The drawback: You lose the ability to perform graphical computation techniques, so you have to rely on vector equations and basic geometry to complete your calculations (which I discuss in Chapter 7).

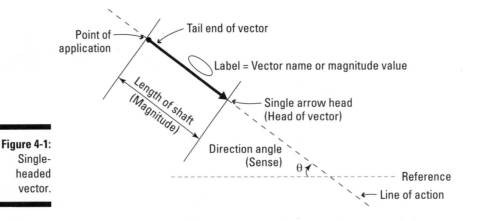

Figure 4-1:
Single-headed
vector.

Beginning in Part V, I show you several different techniques for solving for unknown vector magnitudes. Remember that regardless of which method you use to depict the magnitude, the proper representation of a vector always includes the vector's sense. I delve into the ways you can represent a vector's sense in Chapter 5.

A two-headed monster: The double-headed arrow approach

Another useful notation is the double-headed arrow, which helps depict the rotation of an action such as the turn of a doorknob (see Figure 4-2). The double-headed vector contains much the same information as the single-headed variety. Specifically, the tail, shaft, and label designations of a double-headed vector are all similar to their single-headed cousins described in the preceding section, except this version has two heads.

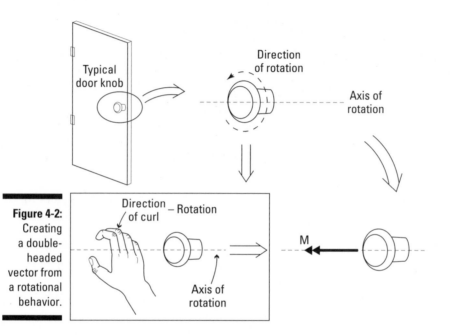

Figure 4-2: Creating a double-headed vector from a rotational behavior.

Single- and double-headed notation are relatively similar; the difference between this illustration and Figure 4-1 is that it has two arrowheads (hence the name) and replaces the line of action with an *axis of rotation.* These changes show that the diagram describes a rotation behavior. The double-headed vector provides some liberties in how you perform calculations with vectors that describe a rotation, which I discuss in more detail in Chapter 12.

Exploring Different Types of Vectors

Most vectors fall into one of three categories, all of which have similar requirements — namely, that they must have a magnitude and a sense. However, not all three categories have a specific point of application. For example, in the case of the sliding vector, the point of action is replaced by a line of action. (Check out the earlier "Defining a Vector" section for more on the required properties of vectors.) Table 4-1 provides you with a snapshot of requirements of the three main vector categories I cover in this section. In the following sections, I explain the different categories of vectors and give examples for each.

Table 4-1	Types of Vectors and Their Requirements		
Type of Vector	*Magnitude*	*Sense*	*Point of Action*
Free	Yes	Yes	No
Fixed	Yes	Yes	Yes
Sliding	Yes	Yes	Not exactly

Fixed vector

A *fixed vector* is a type of vector where the point of application is set at a distinct location and can't be moved without changing the behavior of the initial problem. Examples of fixed vectors include the velocity of a particle and the gravitational weight of a rigid body.

Figure 4-3 illustrates two different fixed vectors. In Figure 4-3a, the fixed vector represents the gravitational influence (or *self weight*) of the object. The *center of gravity* occurs at only one location within an object. (I discuss the center of gravity concept in further detail in Chapter 10.) In Figure 4-3b, the fixed vector represents a velocity vector and indicates the speed and direction of the object. The velocity vector illustrates the behavior of the particle on which it's acting.

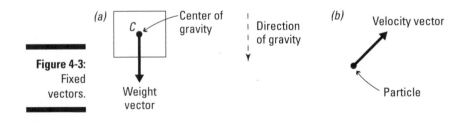

Figure 4-3:
Fixed vectors.

Free vector

Another type of vector is the *free vector,* which doesn't necessarily have a specific point of application but rather acts more generally on an object. These vectors, such as moments and couples (which you can read about in Chapter 11), result in a specific action but may be freely moved around the object without changing the original behavior. Figure 4-4 shows you an example of a free vector. Note how the object rotates with the same intensity (or magnitude) and direction in space regardless of where the action is applied on the object.

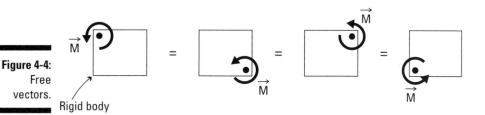

Figure 4-4: Free vectors.

Rigid body

Sliding vector

A third type of vector is the sliding vector, which is also sometimes referred to as a *line vector. Sliding vectors* (such as forces on rigid bodies) may freely move on an object as long as they remain on their line of action. Unlike the fixed vector, the sliding vector doesn't have a distinct point of application but rather acts more generally along a specific direction line (as you can see in Figure 4-5).

In fact, this sliding notion leads to the principle of transmissibility, which is one of the major requirements in the study of rigid body statics (I discuss this principle more in Chapter 9). In statics, a large number of the vectors you deal with are considered sliding vectors, or vectors that have lines of action. Sliding vectors have very useful properties: They can be moved anywhere along their lines of action and still maintain their original behavior. Part III covers these properties in more detail.

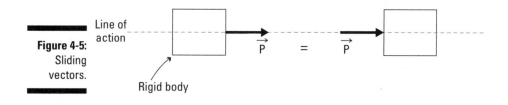

Figure 4-5: Sliding vectors.

Line of action

Rigid body

Chapter 5

Using Vectors to Better Define Direction

. . . .

In This Chapter

▶ Defining direction and Cartesian coordinates

▶ Creating a position vector

▶ Calculating magnitude of a position vector

▶ Developing unit vectors through several different methods

. . . .

*A*fter you have a handle on how to graphically represent all the information required to accurately depict a vector, you're ready to begin tackling different methods for putting vectors to work. The first step is representing the vector in mathematical terms. In this chapter, I describe the notation required to do just that and then show some of the basic calculations that are essential in the vector creation process.

I also show you how to create several basic vectors and even how to use vectors to create additional vectors. All these techniques add more ammunition to your proverbial vector toolbox and are especially essential for simplifying three-dimensional statics problems down the road.

Taking Direction from the Cartesian Coordinate System

The majority of the basic problems that you solve involve the *Cartesian coordinate system,* so the following list introduces you to several important terms related to that system. You can also check out Figure 5-1 for a look at how they work together.

✔ **Axis:** The *axes* are the reference lines that act as a simple ruler for measuring distances of points or objects from a user-defined reference point, known as the *origin*, which I discuss later in this section. In two dimensions, you use two axes: an *x*-axis and a *y*-axis. In three dimensions, you use three: the *x*-, *y*-, and *z*-axes. Each axis indicates a positive and negative direction. You normally only label the positive direction, but you can label both as a reminder.

✔ **Coordinate:** In the Cartesian coordinate system, each point in space is uniquely expressed as a grouping of numerical values called a *coordinate*. For two dimensions, a coordinate is a pair of numerical values written (x,y); in three dimensions, the three values are written (x,y,z).

You measure these coordinates with respect to a base reference point called the origin (see the following bullet). Regardless of how many dimensions you have, the *x*-dimension is always the distance from the origin measured parallel to the *x*-axis. Similarly, the *y*- and *z*-dimensions are measured parallel to the *y*- and *z*-axes, respectively. Coordinates may be either positive or negative, indicating their relation to the origin.

✔ **Origin:** The *origin* is a very special point at which all the axes intersect each other. The coordinates of the origin are traditionally taken as (0,0) or (0,0,0), and that's what I assume in this book.

✔ **Scale:** The *scale* of your Cartesian representation indicates what units you're measuring in. The individual units of the Cartesian coordinate system are completely up to you to decide, but remember to be mindful of significant digits and accuracy (which I cover in Chapter 2). You can measure every distance in cosmic light years if you want, but most distances here on Earth are normally measured in feet and inches (U.S. customary units) or meters (SI/metric units). In many problems, using these units lets you minimize problems with numerical accuracy and significant digits.

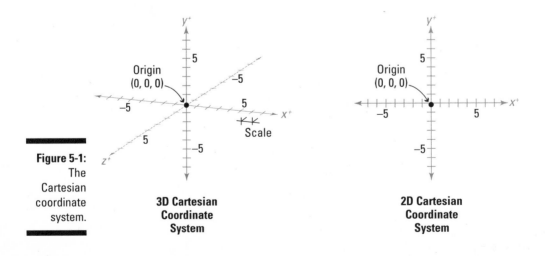

Figure 5-1:
The
Cartesian
coordinate
system.

**3D Cartesian
Coordinate
System**

**2D Cartesian
Coordinate
System**

As a Crow Flies: Using Position Vectors to Determine Direction

To start the vector creation process, I present you with the *position vector,* a simple vector that describes how to get from Point A to Point B. The position vector defines the most direct path from one point to another. In statics, you can even use position vectors to construct other types of vectors, such as unit vectors, which I show later in this chapter.

Unlike normal driving directions, which may have turns, detours, and even backtracking associated with them, a position vector is concerned only with the most direct path from one point to another. Imagine a hiker walking along a winding trail; he may zig and zag in many directions as he hikes toward his destination. However, this path often isn't the most direct route possible. The true path is often limited by physical driving or walking considerations such as availability of roads and bridges or the simple fact that most humans can't fly.

Consider the navigation example in Figure 5-2. Suppose you're standing at Point A, and your final destination is the top of a nearby hill, designated as Point B. Point A and Point B both have unique coordinates in space (otherwise, it would be a really short trip). The figure already includes a three-dimensional Cartesian coordinate system and indicates the scale, origin, units, and positive and negative directions for the scale. Clearly, the most direct route you can take is along a straight line that connects Point A with Point B, or the proverbial path "as a crow flies." A crow, or any bird for that matter, isn't subject to the roads or constraints humans are, so it's free to simply focus on getting from Point A to Point B. In statics, position vectors let you do the same thing.

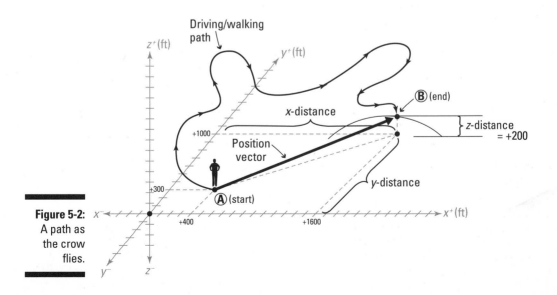

Figure 5-2: A path as the crow flies.

Describing direction in detail

The position vector retains all of the regular properties of a vector that I discuss in Chapter 4 in that it still has a *magnitude* (length), *sense* (direction), and *point of application* (location in space).

Most texts denote the position vector with a lower case r with subscripts denoting, in order, the start and stopping points of the vector. A position vector from Point A to Point B would be labeled $\mathbf{r}_{AB}$. Conversely, the position vector that describes the reverse direction, starting at Point B and ending at Point A would be labeled $\mathbf{r}_{BA}$. Although the two position vectors are connecting the same two points, these vectors are uniquely different, as I discuss in the following section.

Figure 5-3 shows the vector from the example in the preceding section and helps demonstrate the representation used to define a position vector.

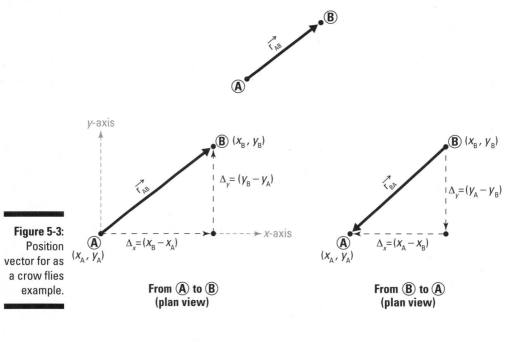

Figure 5-3:
Position
vector for as
a crow flies
example.

Moving from Point A to Point B and back again

By establishing the locations of Point A and Point B, you can define the Cartesian coordinates for these points. The order of the start point (vector tail) and stop point (vector head) is extremely important when you're creating a position vector. The first step is to establish a coordinate system if one isn't already declared. In the Figure 5-2 example, I assume that a positive Cartesian *x*-direction is to the east, and a positive Cartesian *y*-direction is to the north. A positive elevation is vertical upwards from the *xy* plane.

The following formulas show you how you can easily find the relative distance traveled along a line between the start and end points.

- *x*-distance traveled from start to end: $\Delta_x = (x_{END} - x_{START})$
- *y*-distance traveled from start to end: $\Delta_y = (y_{END} - y_{START})$

To find the relative distance traveled from end to start, simply reverse the x_{END} and x_{START} terms in the previous calculations. In the event that your destination is at a different elevation (such as being on a hill or in a valley) than your starting point, you need to include one extra dimension, the Cartesian *z*-direction — you just substitute the *z* values: $\Delta_z = (z_{END} - z_{START})$

In this direction, a positive change in the Cartesian *z*-direction indicates that the end point is above the start point. Likewise, a negative value indicates that the end point of the vector is below the start point. These generic relationships easily take into account positive and negative distance values.

If you're careful with the math, the signs of the distances can actually help you with the sense of the vector.

In the first part of Figure 5-3 earlier in the chapter, you can see that Point B is located northeast of Point A. You can define the Cartesian coordinates for Point A and Point B from the dimensions and units indicated:

- Point A = (x_A, y_A, z_A) = (+400,+300,0)
- Point B = (x_B, y_B, z_B) = (+1600,+1000,+200)

where (x,y,z) are the coordinates of their respective points. The distance that needs to be traveled to the right (or east, or positive) is given by the distance $(x_B - x_A)$, and the distance north would be given by the distance $(y_B - y_A)$. More generally:

- *x*-distance traveled from Point A to Point B: $\Delta_x = (x_B - x_A) =$ (+1,600 − (+400)) = +1,200 feet
- *y*-distance traveled from Point A to Point B: $\Delta_y = (y_B - y_A) =$ (1,000 − (+300)) = +700 feet
- *z*-distance traveled from Point A to Point B: $\Delta_z = (z_B - z_A) =$ (+200 − 0) = +200 feet

If you travel in the opposite direction from Point B to Point A (refer to the second part of Figure 5-3), you get the same numeric values, but the signs in front are now different (-1,200 feet, -700 feet, and -200 feet).

A First Glance at Determining a Vector's Magnitude

After you have a handle on the three basic properties of a vector (magnitude, sense, and point of application) from Chapter 4 and the information to display them accurately, you're ready to start making calculations. The following sections investigate the calculations for magnitude.

Recognizing the notation for magnitude

In practice, you encounter the magnitude of a random vector **F** denoted as $\left\|\vec{F}\right\|$, where the vector name is bound by the *norm* designation (or the double vertical lines). In many engineering mechanics books, you encounter a simpler notation depicted as $\left|\vec{F}\right|$ with single absolute value brackets, especially in equations. I use italics to indicate magnitude in the text. As I note in the Introduction, these last two methods are the notation that I use in this text.

Computing the magnitude of a position vector: Pythagoras to the rescue!

Pythagoras of Samos was a Greek philosopher and mathematician (circa 570 B.C.) who is credited with discovering the Pythagorean theorem, which proves the relationship between the sides of a right (or 90-degree) triangle (see Figure 5-4). To calculate the magnitude of a vector, you need to use your ability to locate right triangles and apply some basic geometry.

Figure 5-4:
The
Pythagorean
theorem.

The two-dimensional Pythagorean theorem

Say you have a position vector between two points. Point A has coordinates of (+400,+300,0) and Point B has coordinates of (+1600,+1000,+200). (If this sounds like the example in Figure 5-2 earlier in the chapter, that's because it is.) The magnitude of this position vector is actually the direct distance between Point A and Point B. For a simple two-dimensional problem, you can calculate the distance between those two points by employing the two-dimensional Pythagorean theorem ($C^2 = A^2 + B^2$; see Chapter 2).

Going vertical: The Pythagorean theorem in three dimensions

For a three dimensional problem, you need to do a bit more calculation but you can still employ the Pythagorean theorem by simply constructing two right triangles inside a box of known dimensions A, B, and C as shown in Figure 5-5.

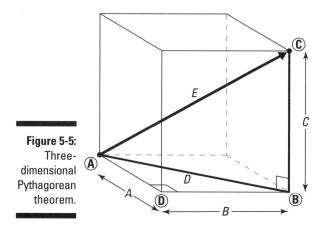

Figure 5-5:
Three-dimensional Pythagorean theorem.

The distance between two points at different elevations is equal to the square root of the sum of the squares of the sides of a right-angle box that fully contains the start and end points on opposite corners.

For example, in Figure 5-5, D is the hypotenuse of the first right triangle, ABD. The Pythagorean theorem tells you that for triangle ABD, $D^2 = A^2 + B^2$

Similarly, you can also create a second, vertically-oriented right triangle ABC with a hypotenuse of E and sides C (the height of the box) and D (the previously calculated hypotenuse of the first triangle). Using the Pythagorean theorem yet again, you can state that $E^2 = C^2 + D^2$. You can then substitute the equation for triangle ABD (which represents the value of D^2) into this equation to get $E^2 = A^2 + B^2 + C^2$ or

$$E = \sqrt{\left(A^2 + B^2 + C^2\right)}$$

Putting Pythagoras to work

In the case of the navigation example in Figure 5-2 earlier in the chapter, which shows a position vector between Points A and B, the box that contains these two points has sides of distances Δ_x, Δ_y, and Δ_z. Substituting these values into the final equation in the preceding section allows you to compute the magnitude of the position vector between Point A and Point B:

$$\text{Distance} = |\vec{r}| = \sqrt{\left(\left(\Delta_x\right)^2 + \left(\Delta_y\right)^2 + \left(\Delta_z\right)^2\right)}$$

Fortunately, the magnitude of a position vector has a specific meaning that you can actually see and measure in that it precisely represents the direct distance between those two points.

Figure 5-6 illustrates the three-dimensional box required to compute the magnitude of the position vector between Point A and Point B.

$$\text{Distance} = \left|\overrightarrow{r_{AB}}\right| = \sqrt{\left(\left(1{,}600 - \left(400\right)\right)^2 + \left(1{,}000 - \left(300\right)\right)^2 + \left(200 - \left(0\right)\right)^2\right)}$$
$$= 1403.56 \text{ ft}$$

The distance calculated is the magnitude of the position vector $\mathbf{r_{AB}}$ starting at Point A and ending at Point B.

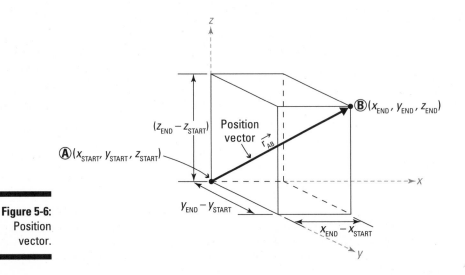

Figure 5-6:
Position
vector.

One final note on position vectors: Although the distance between Point A and Point B is a specific value, the vector $\mathbf{r_{AB}}$ that defines the path from Point A to Point B isn't the same as the position vector $\mathbf{r_{BA}}$ from Point B to Point A. That is, the magnitudes are the same $\left(\left|\overrightarrow{r_{AB}}\right| = \left|\overrightarrow{r_{BA}}\right|\right)$, but the vectors are different $\left(\overrightarrow{r_{AB}} \neq \overrightarrow{r_{BA}}\right)$.

Unit Vectors Tell Direction, Too!

Although it's not exactly a position vector (see the preceding section), a *unit vector* is also used frequently to help describe directions between points in space, particularly vectors' lines of action. Unit vectors prove to be extremely useful in the construction of other vectors, particularly force vectors, which I cover more in Chapter 9.

The main difference between the position vector and a unit vector is that the position vector tells precisely how to get from one point in space to another and the magnitude of the position vector is actually the physical distance between the two points. The unit vector, on the other hand, is a vector used for describing the orientation of a line that passes through those two points. So although a unit vector isn't as useful for calculating distance, it proves to be very handy for specifying direction. In a way, you can think of a unit vector as your finger pointing at your final destination as you stand at the starting point.

Cartesian-vector notation

In this book, I denote a unit vector by the label **u** and attach a label as a subscript to help describe the direction. Just as before, I denote the magnitude of a unit vector as $|\vec{u}|$ and attach subscripts to this notation to help define the direction, relative to two points on the line of action. This line is actually the same line as the line of action of the position vector.

Unlike the magnitude of the position vector, the magnitude of a unit vector is always exactly one unit long. That is,

$$|\vec{u}| = 1.0$$

Whenever you use the Cartesian coordinate system (see "Taking Direction from the Cartesian Coordinate System" earlier in the chapter), you can make use of three very special unit vectors. As shown in Figure 5-7, you can define a vector of magnitude 1.0 in the positive direction for each of the principle Cartesian axes.

- ✔ **x-direction:** The unit vector parallel to the *x*-axis has the designation of a bolded **i** or a special marker (kind of like a party hat) over the unbolded letter as in the following: $\vec{u_x} = \hat{i}$

- ✔ **y-direction:** The unit vector parallel to the *y*-axis has the designation of a bolded **j** or the special hat marker over the unbolded letter: $\vec{u_y} = \hat{j}$

- ✔ **z-direction:** The unit vector parallel to the *z*-axis has the designation of a bolded **k** or the special marker over the unbolded letter: $\vec{u_z} = \hat{k}$

The arrow over the vector label **u**$_z$ is the same as bolding; it's just another way to designate that you're talking about a vector.

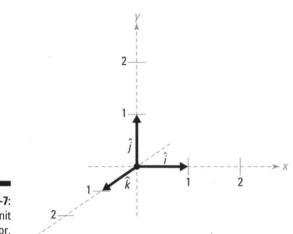

Figure 5-7:
A unit
vector.

Using unit vectors to create position vectors

To write a vector in Cartesian vector notation, you take full advantage of those three unit vectors (**i, j,** and **k**) that I talk about in the preceding section. For example, if you have a vector **V** with magnitude $\left|V_x\right|$ of 10 meters in the positive x-direction, you can write this vector as

$$\vec{V} = 10\hat{i} + 0\hat{j} + 0\hat{k} \text{ meters}$$

You can also have a vector going in a combination of directions. Suppose that vector **V** has a piece in the x-direction of magnitude $\left|\overrightarrow{V_x}\right|$ that equals 15 meters and a piece in the negative z-direction of magnitude $\left|\overrightarrow{V_z}\right|$ that equals 26 meters. You can then write the vector **V** as

$$\overline{V} = 15\hat{i} + 0\hat{j} - 26\hat{k} \text{ meters}$$

Because there's no y-direction component, the coefficient in front of the **j** (y-direction) is zero. You also notice that the coefficient in front of the **k** (z-direction) is a negative value. This negative value simply indicates that the magnitude of that piece of the vector is acting in the opposite direction from the assumed positive z-direction.

Position vectors can be Cartesian too!

In terms of the generic coordinates (x_A, y_A, z_A) for Point A and coordinates (x_B, y_B, z_B) for Point B, you can write the position vector from Point A to Point B as

$$\overrightarrow{r_{AB}} = \Delta_x \hat{i} + \Delta_y \hat{j} + \Delta_z \hat{k}$$
$$= (x_B - x_A)\hat{i} + (y_B - y_A)\hat{j} + (z_B - z_A)\hat{k}$$

or even more generically as

$$\overrightarrow{r_{AB}} = (x_{END} - x_{START})\hat{i} + (y_{END} - y_{START})\hat{j} + (z_{END} - z_{START})\hat{k}$$

As long as you choose the points for the start and the end of the vector correctly, and you can correctly determine the Cartesian coordinates of each of those points, the signs of the coefficients in front of each of the unit vectors **i**, **j**, and **k** take care of themselves. In fact, the signs of the *scalar* values (dealing only with magnitude and not with sense) are what help you determine the sense of the vector — a negative scalar coefficient tells you that piece of the vector is acting in the negative direction of the unit vector noted immediately after that scalar value.

For example, try to write the actual position vector in Cartesian coordinates from Point A to Point B for a navigation example where Point A has coordinates of (+400,+300,0) and Point B has coordinates of (+1600,+1000,+200). (For a visual, check out Figure 5-2 earlier in the chapter.) The notation for the label of this vector is **r**$_{AB}$.

In this example, the change in the *x*-dimension (Δ_x) is 1,200 feet, the change in the *y*-direction (Δ_y) is 700 feet, and the change in the *z*-direction (Δ_z) is 200 feet. (To see the calculations that produce these figures, check out "Moving from Point A to Point B and back again" earlier in the chapter.)

You can now write the position vector from Point A to Point B as

$$\overrightarrow{r_{AB}} = 1,200\hat{i} + 700\hat{j} + 200\hat{k} \text{ feet}$$

Relationship between a vector, its magnitude, and its direction

In the example in the preceding section, you create a simple position vector **V** = 10**i** + 0**j** + 0**k** meters by using a known distance (the magnitude) and its direction. You can simplify this vector even further by omitting the terms that have zero as their coefficients. This step leaves a new form of the vector **V**:

$$\overline{V} = 10\hat{i} \text{ meters}$$

The original vector **V** was created from the given information that the magnitude of the vector is 10 meters parallel to the positive Cartesian x-axis. If you examine this setup in general terms, you may conclude the following relationship:

$$\text{vector} = \text{magnitude} \cdot \text{direction}$$

For the navigation example, you can substitute the following terms into the equation:

- vector : $\overline{V}$
- magnitude : $\left|\overline{V}\right| = 10$ meters
- direction : unit vector in the positive Cartesian x-direction

$$: \overrightarrow{u_V} = \hat{i}$$

Thus, the expression above is more accurately written as:

$$\overline{V} = \left|\overline{V}\right| \cdot \overrightarrow{u_V}$$

Notice that this expression directly relates the vector itself as the multiplication of its scalar magnitude and a unit vector in the direction of that magnitude. Imagine that! Creating a new vector from a unit vector!

Creating Unit Vectors from Scratch

In the preceding section, I identify a unit vector as a means to define the direction of a vector's line of action. In this section, I explain several basic techniques for creating a unit vector. After you have this step accomplished, you can easily construct a properly defined vector notation for any action.

Shrinking down position vectors

In the section "Relationship between a vector, its magnitude, and its direction" earlier in the chapter, I develop the expression for a vector based on its magnitude and direction. If you do a little rearranging of the last equation of that section, you get the following expression:

$$\overrightarrow{u_V} = \frac{\overline{V}}{\left|\overline{V}\right|} = \frac{\text{Vector, V}}{\text{Magnitude of Vector, V}}$$

This expression illustrates that you can establish the direction of a vector's line of action by taking the vector representation over the magnitude of that same vector. Suppose you have a position vector from Point A to Point B given as

$$\overrightarrow{r_{AB}} = 1{,}200\hat{i} + 700\hat{j} + 200\hat{k} \text{ feet}$$

with a magnitude of

$$\left|\overrightarrow{r_{AB}}\right| = 1{,}403.56 \text{ feet}$$

(You may recognize these numbers from the example in "Unit Vectors Tell Direction, Too!" earlier in the chapter.) You can easily create a unit vector to describe the orientation of the line of action between Point A and Point B (the line that connects those two points):

$$\overrightarrow{u_{AB}} = \frac{1{,}200\hat{i} + 700\hat{j} + 200\hat{k} \text{ feet}}{1{,}403.56 \text{ feet}}$$

$$= \frac{1{,}200i \text{ feet}}{1{,}403.56 \text{ feet}} + \frac{700\hat{j} \text{ feet}}{1{,}403.56 \text{ feet}} + \frac{200\hat{k} \text{ feet}}{1{,}403.56 \text{ feet}} = 0.854\hat{i} + 0.499\hat{j} + 0.142\hat{k}$$

Notice that a unit vector has no actual units because the vector terms have units of feet in the numerator of each term, and the magnitude (or distance) gives units of feet in the denominator, so they cancel out. To verify that this vector meets the criteria of being a unit vector, you simply need to calculate its magnitude:

$$\left|\overrightarrow{u_{AB}}\right| = \sqrt{(0.854)^2 + (0.499)^2 + (0.142)^2} = 0.999 \approx 1.000$$

Realize that because the coefficients of the unit vector terms are typically irrational numbers, the magnitude rarely computes to be exactly 1.0. This discrepancy is another example of the importance of significant digits and computational accuracy that I introduce in Chapter 2. For the purposes of the example here, you'll be happy that 0.999 is approximately 1.000.

Using angular data and direction cosines

Another technique that's sometimes handy for creating a unit vector utilizes calculations involving the direction cosines. The *direction cosines* represent the angles between any two given vectors — even two unit vectors! You can use the three Cartesian axes as the reference because you already have those three special **i, j,** and **k** unit vectors to define them. (Flip to the earlier section "Cartesian vector notation" for the details on these unit vectors.)

One of the major difficulties of using direction cosines is in actually determining the angle between the vector or line of action of interest and the Cartesian axis. These angles often occur on geometric planes that aren't perpendicular to any Cartesian axis and therefore may be difficult or cumbersome to compute.

Figure 5-7 earlier in the chapter illustrates the three unit vectors with respect to each of the three principal Cartesian axes. For example, Figure 5-8a shows the *x*-direction unit vector $\mathbf{u}_x = \mathbf{i}.$ The direction cosine for this vector, the angle α, is the angle between the line of action of the vector you're working with and the line of action of the *x*-direction unit vector, contained within the plane of those two vectors. Similarly, Figure 5-8b shows the *y*-direction unit vector $\mathbf{u}_y = \mathbf{j}.$ The direction cosine for this vector, the angle β, is the angle between the line of action of your vector and the *y*-direction unit vector. Figure 5-8c shows the *z*-direction unit vector $\mathbf{u}_z = \mathbf{k},$ and the direction cosine for this vector, the angle γ, is the angle between the line of action of your vector and the *z*-direction unit vector, contained within the plane of the two vectors.

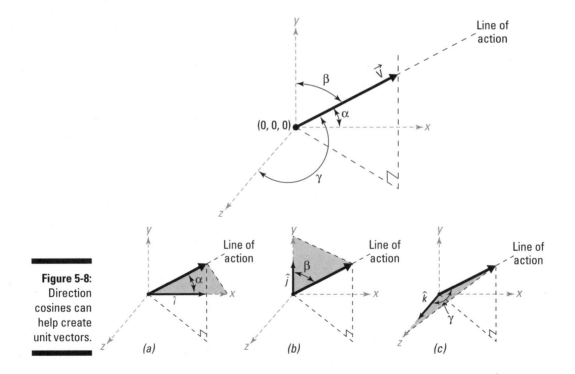

Figure 5-8: Direction cosines can help create unit vectors.

Note: In some texts, you may see the angles for the direction cosines referred to as θ_x, θ_y, and θ_z to represent the angle between the vector's line of action and its corresponding Cartesian axis unit vector. Here I use α, β, and γ, respectively.

The previous example showed how to piece together different parts of a vector in order to compute a different vector. You can use a similar technique with direction cosines to create a unit vector as follows: $\vec{u}_V = (\cos\alpha)\hat{i} + (\cos\beta)\hat{j} + (\cos\gamma)\hat{k}$. The only information you need for this example is the three angles between the line of action and each of the principal Cartesian axes.

Utilizing proportions and similar triangles

Yet another method that you may find useful for creating unit vectors utilizes the proportions or ratios of the dimensions of a vector's or object's line of action.

In some textbooks, you see a vector or line of action with an odd-looking pair of numbers written around a triangle. In the case of the vector, this small triangle is attached to the shaft of the vector. For a line of action, you see the triangle attached somewhere along the line, or you can infer it from the given dimensions of an object or distance between given points. This proportion technique is useful because you eliminate the need to even calculate the angle for the line of action, and as a result you may actually improve issues with accuracy and significant digits, as I mention in Chapter 2.

This *proportion triangle* represents a horizontal and vertical proportion and is comparable to the slope diagrams you may remember from your algebra class. The horizontal line segment of this proportion triangle represents the horizontal proportion, and similarly the vertical line segment represents the vertical proportion.

The secret to using these proportion values lies in applying the basic trigonometry functions sine, cosine, and tangent (which I touch on in Chapter 2). In Figure 5-9, each of the three proportion triangles shown all have the same angle θ in common. If you draw a right triangle such as the first one shown, you can easily compute θ from

$$\tan\theta = \frac{\text{vertical proportion}}{\text{horizontal proportion}}$$

You can then take that angle and plug it back into the direction cosines formula from the preceding section.

For the example of Figure 5-9, you can compute the angle between the horizontal x-axis and the vector is

$$\theta_x = \tan^{-1}\left(\frac{\text{vertical proportion}}{\text{horizontal proportion}}\right) = \tan^{-1}\left(\frac{4}{-3}\right) = -53.13°$$

Note that the horizontal proportion in this example is a negative value because it's measured in the direction opposite to the positive *x*-axis. The vertical proportion is positive because it's measured in the same direction as the positive *y*-axis.

In this example, there's no *z*-dimension. In fact, you rarely see proportional dimensions in three dimensions, due to the difficulty of clearly representing the proper values in a drawing.

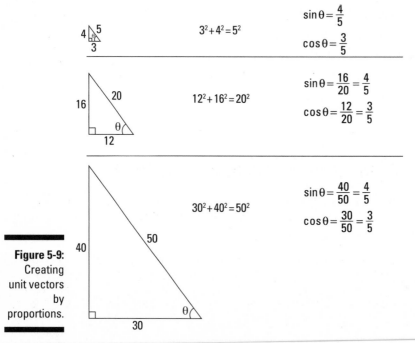

Figure 5-9:
Creating
unit vectors
by
proportions.

As drawn, the angle θ is a direction cosine with respect to the *x*-axis. Similarly, you can calculate the remaining direction cosines for the other axes and plug them straight into the unit vector notation from the preceding section. Just remember, by first calculating this angle, you normally end up dealing with an irrational numerical value when you apply a sine, cosine, or tangent function.

In Figure 5-9, the proportion triangle is actually a right triangle. Using the Pythagorean theorem (which I discuss earlier in this chapter), you can quickly compute its hypotenuse:

$$\text{hypotenuse} = \sqrt{(-3)^2 + (4)^2} = 5$$

You can then calculate the direction cosine with respect to the positive *x*-axis as

$$\cos\theta_x = \cos\alpha = \frac{-3}{5} = -0.6$$

and the direction cosine with respect to the positive *y*-axis as

$$\cos\theta_y = \cos\beta = \frac{4}{5} = +0.8$$

Note that for the third dimension, the direction cosine for the *z*-axis is 90 degrees, and consequently cos(90) = 0.

Assembling these into the unit vector equation as follows:

$$\vec{u} = (\cos\alpha)\hat{i} + (\cos\beta)\hat{j} + (\cos\gamma)\hat{k}$$

produces a unit vector

$$\vec{u} = -0.6\hat{i} + 0.8\hat{j} + 0\hat{k}$$

Knowing which technique to use

So how do you know which of the techniques in the preceding sections is best for your situation? The following checklist shows you some factors that affect your choice:

✔ **Position vectors:** Position vectors are most readily utilized in problems that can be defined by Cartesian coordinates. For example, problems involving points on a map and lengths of cables or ropes are all candidates for using position vectors. Problems in three dimensions often use position vectors in one manner or another. In fact, a three-dimensional problem is a very strong indicator that position vectors may be worth checking into.

✔ **Direction cosines:** Direction cosines aren't as common as problems utilizing Cartesian coordinates. However, if a problem doesn't provide any linear dimension data, that may be a good indicator that you need a direction cosine calculation.

You need all three angles to be able to create your unit vector. If you can't find all three angles, you can't use the direction cosine method.

✔ **Proportions and similar triangles:** Proportions and similar triangles can be a bit easier to spot. Namely, you can manipulate problems with a vector or line of action that have the proportion triangle directly denoted with this technique. One major reason for using this method is a lack of suitable information (such as angular information or Cartesian coordinate data) to use either of the other methods.

Chapter 6

Vector Mathematics and Identities

• •

• •

*A*fter you're familiar with how to depict the velocity of a particle (see Chapter 4) and write the equation for a force vector (see Chapter 5), you want to start looking at how to work with those vectors through mathematics. Vectors become especially important when you work three-dimensional statics problems, and the skills I show you in this chapter introduce some of the methods for performing calculations with them.

In working with vectors, you soon discover a variety of basic operations that are similar to many of the basic mathematic operations you have used when working with scalar values. However, there are also some special rules that you need to observe. In this chapter, I introduce these operations and rules and show you how to apply them to your vector problems. I also give you a convenient list of properties you'll use with these operations throughout your statics work.

Performing Basic Vector Operations

As you may have learned in your conventional math classes, addition and subtraction are among the most basic (and important) calculations that you work with. Vectors are no different; addition, subtraction, and relocation all become important skills, and that's what I cover in the following sections.

Adding vectors

Simply put, the addition of vectors involves collecting each of the pieces of the action that are acting in a common direction and then representing them with some indicator of the direction of those pieces. This indicator can be another vector and in fact is often a *unit vector* (which is a special type of vector that has a magnitude of exactly 1.0) in the direction of each of the three *Cartesian axes* (covered in Chapter 5).

To add two vectors together, you simply add the scalar coefficients in front of each unit vector (**i, j,** and **k**) to make new scalar coefficients. Consider two vectors, $\mathbf{P_1}$ and $\mathbf{P_2}$:

$$\vec{P_1} = -4\hat{i} + 10\hat{j} - 23.2\hat{k}$$

and

$$\vec{P_2} = 10\hat{i} - 13.2\hat{j} - 26\hat{k}$$

The sum of these two is then

$$\vec{P}_{TOTAL} = \vec{P_1} + \vec{P_2} = (-4+10)\hat{i} + (10 + (-13.2))\hat{j} + (-23.2 + (-26))\hat{k}$$
$$= 6\hat{i} - 3.2\hat{j} - 49.2\hat{k}$$

To further illustrate this concept, use Figure 6-1 to define the following navigation problem.

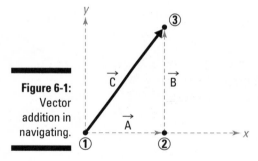

Figure 6-1:
Vector
addition in
navigating.

Using the construction techniques I discuss in Chapter 5, you can see that the position vector that defines the direct path from Point 1 to Point 3 can be given by

$$\vec{C} = \vec{r_{13}} = 40\hat{i} + 80\hat{j}$$

Instead of creating a position vector directly between Point 1 and Point 3, suppose you want a more roundabout path. You define a position vector **A** for path #1 between Point 1 and Point 2, such that

$$\vec{A} = 40\hat{i}$$

and another position vector **B** for path #2 from Point 2 to Point 3, such that

$$\vec{B} = 80\hat{j}$$

You can also travel a different path but still reach the same destination. That is, you can start at Point 1 and travel directly to Point 2 (along the position vector **A**), and then turn and travel from Point 2 to Point 3 (along the position vector **B**). In this case, your start point and end point would be exactly the same as the direct path. When you write this path out, you can see that the new path has the same start point and end point and is simply the sum of the individual path segments of the two legs of the trip (**A** and **B**). In mathematical terms,

$$\vec{r_{13}} = \vec{r_{12}} + \vec{r_{23}} = \vec{A} + \vec{B} = 40\hat{i} + 80\hat{j}$$

Substituting in the expressions for **A** and **B**, notice that the result is the exact same vector. This simple example illustrates the concept of addition of vectors. What you may notice is that the original, **C,** is actually the sum of the individual paths (**A** and **B**) that are taken, and the ordering of the paths does not matter.

Check out the later section "Useful Vector Operation Identities" for some handy vector addition properties.

Subtracting vectors

Subtracting vectors is basically the same operation as adding vectors (see the preceding section), only in reverse. The only difference is that you actually convert the vector being subtracted to a negative vector and then add the vectors. To create a negative vector, you just need to reverse the signs of each of the scalar coefficients; you can do so by simply multiplying each of the scalar terms by –1. For example, look at the following vector P_1:

$$\vec{P_1} = -4\hat{i} + 10\hat{j} - 23.2\hat{k}$$

The negative vector of this vector is

$$-\vec{P_1} = +4\hat{i} - 10\hat{j} + 23.2\hat{k}$$

If you want to subtract vector P_1 from vector P_2 in the preceding section, you just add the negative of vector P_1 to vector P_2 as shown in the following equation:

$$\vec{P_2} - \vec{P_1} = \vec{P_2} + \left(-\vec{P_1}\right) = \left(10 + (+4)\right)\hat{i} + \left(-13.2 + (-10)\right)\hat{j} + \left(-26 + (+23.2)\right)\hat{k}$$
$$= 14\hat{i} - 23.2\hat{j} - 2.8\hat{k}$$

Figure 6-2 illustrates the subtraction of two vectors. Notice that the final result of the operation is an entirely different vector from the vector created by addition.

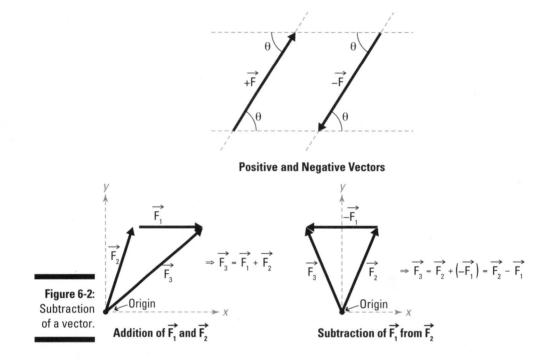

Positive and Negative Vectors

Figure 6-2:
Subtraction
of a vector.

Addition of $\vec{F_1}$ and $\vec{F_2}$

$\Rightarrow \vec{F_3} = \vec{F_1} + \vec{F_2}$

Subtraction of $\vec{F_1}$ from $\vec{F_2}$

$\Rightarrow \vec{F_3} = \vec{F_2} + \left(-\vec{F_1}\right) = \vec{F_2} - \vec{F_1}$

In this regard, the subtraction of a vector is similar to the subtraction of two scalar quantities: The term being subtracted is simply the addition of its negative representation.

Moving vectors head to tail

In the earlier sections in this chapter, I illustrate the basic vector concepts as a series of simple steps: First, action #1 occurs, followed by another action #2. However, in the physical world, this sequence may or may not be the case. In fact, a vector may experience multiple actions simultaneously. Moving vectors head to tail when adding them is a quick and easy way of working with simultaneous actions. In fact, if you have a hundred simulta- neous actions on an object, connecting the tail of one action to the head of another action for every action on the object helps you determine the com- bined response. The final combined response will be from the tail of the very first action you listed to the head of the very last action you listed.

Order doesn't matter for simultaneous events. However, you can only attach a single tail of a vector to any given vector arrowhead. You can't attach the tails of multiple vectors to the head of the same vector.

For example, Figure 6-3a shows a baseball that has been struck by a bat. The baseball may experience a velocity in both the upward direction (as defined by vector **B**) as well as a velocity in a horizontal direction (as defined by vector **A**). Velocity **A** makes the ball move away from the batter, and velocity **B** makes it rise in the air. Each of these actions is independent of the other, and each may have a significantly different magnitude of action.

Figure 6-3:
Simulta-
neous
actions on a
baseball.

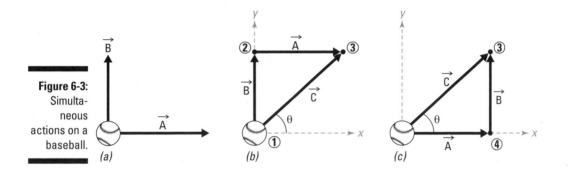

As I discuss in the "Adding vectors" section earlier in the chapter, you can represent the combined action on the baseball by adding the combined actions on the ball.

Figure 6-3b illustrates the case where the action **B** is drawn first. At the conclusion of action vector **B** (or at its head), the action **A** begins acting in its own direction. That is, at the conclusion of action **B,** the tail of action **A** begins.

Similarly, in Figure 6-3c, action **A** can be the first action that affects the baseball. Upon the conclusion of action **A,** action **B** begins. Thus, the tail for the **B** action is attached to the head of **A.**

Suppose you define a new vector **C** as being the combined action of **A** and **B.** The new vector **C** shown in both scenarios of Figure 6-3b and c, results in the magnitude, sense, and angle for the line of action (θ) of **C** being identical. If the three major properties of a vector are the same, the two vectors are actually the same. (Flip to Chapter 4 for more info on these basic vector properties.)

What Do You Mean I Can't Multiply Vectors? Creating Products

Mathematically speaking, adding and subtracting vectors and scalars are basically the same operation. However, the two remaining scalar operations — multiplication and division — work a little differently with vectors.

You can't directly multiply two vectors, but you do have other unique operations at your disposal, such as *products.* The following sections deal with two of the more popular products: dot products and cross products.

"Useful Vector Operation Identities" later in the chapter shows you some properties to keep in mind as you work with products.

Dot products

The *dot product* is a type of operation that allows you to create a *projection,* or the portion of one vector that acts in the same direction as a second vector; it always produces a scalar result. It just requires knowing the magnitude of the two vectors involved and the angle between their lines of action. This type of operation proves to be useful in physics calculations and for quickly determining the action of one vector along the line of action of another vector. After all, if you've gone through all of the trouble of specifying the direction of a vector in its notation, it only makes sense that you should be able to put that information to work as well.

Figure 6-4 illustrates two different vectors **A** and **B** that are oriented at some arbitrary angle, θ, between them.

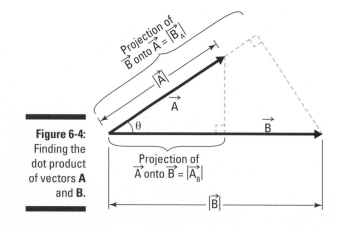

Figure 6-4:
Finding the dot product of vectors **A** and **B**.

The dot product of Figure 6-4's vectors is defined as follows:

$$\vec{A}\cdot\vec{B} = \left(\left|\vec{A_B}\right|\right)\left|\vec{B}\right| = \left(\left|\vec{B_A}\right|\right)\left|\vec{A}\right|$$
$$= \left(\left|\vec{A}\right|\cos\theta\right)\left|\vec{B}\right| = \left(\left|\vec{B}\right|\cos\theta\right)\left|\vec{A}\right|$$
$$= \left|\vec{A}\right|\left|\vec{B}\right|\cos\theta$$

In these equations, $\left|\vec{A_B}\right|$ represents the portion of vector **A** that's acting in the direction of vector **B.** Similarly, $\left|\vec{B_A}\right|$ represents the portion of vector **B** that's acting in the direction of vector **A.**

Cross products

The *cross product* is an operation performed on two different vectors that produces a third vector that is orthogonal (perpendicular) to each of the original vectors.

Don't confuse the cross product operator (×) with the x-style multiplication operator × you learned early in your math career. These are distinctly different operations.

The cross product proves to be very useful in calculating rotational quantities called *moments,* which I cover in Chapter 12.

Unlike the dot product, which returns a scalar quantity, the cross product computation always returns a new vector (complete with Cartesian vector notation). Check out the preceding section for dot product details.

Figure 6-5 illustrates two different vectors **A** and **B** that are oriented at some arbitrary angle θ between them.

The cross product for Figure 6-5 is defined as follows:

$$\vec{A}\times\vec{B} = \left|\vec{A}\right|\left|\vec{B}\right|\sin\theta\cdot\vec{n}$$

where **n** is a normal vector to both **A** and **B.** The challenge in calculating the cross product is usually in calculating the normal vector **n.** If you know the direction of **n,** the computation isn't much more difficult than the dot product calculation in the preceding section.

Unfortunately, that same vector **n** is often an unknown entity. Fortunately, there is a second identity, involving a *determinant* (a mathematical operation that utilizes a 3-x-3 matrix of values) that is much easier to calculate. The first line of the determinant always contains the unit vectors in the direction of each of the Cartesian axes. The second line is always the scalar coefficients of each of the unit vectors for the first vector listed in the cross product. The third line is always the scalar coefficients of each of the unit vectors for the second vector listed in the cross product.

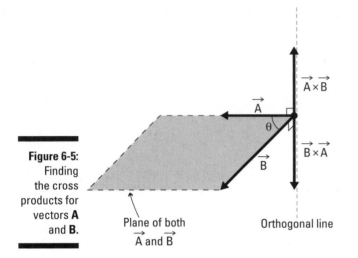

Figure 6-5:
Finding the cross products for vectors **A** and **B**.

Plane of both $\vec{A}$ and $\vec{B}$

Orthogonal line

For example, say you have a vector **A** defined as

$$\vec{A} = A_x\hat{i} + A_y\hat{j} + A_z\hat{k}$$

where A_x, A_y, and A_z are scalar components in the Cartesian x-, y-, and z-directions respectively, and a vector **B** defined as

$$\vec{B} = B_x\hat{i} + B_y\hat{j} + B_z\hat{k}$$

where B_x, B_y, and B_z are scalar components in the Cartesian x-, y-, and z-directions respectively. You can then calculate the cross product from the determinant by using the following setup

$$\vec{A} \times \vec{B} = \begin{vmatrix} \hat{i} & \hat{j} & \hat{k} \\ A_x & A_y & A_z \\ B_x & B_y & B_z \end{vmatrix}$$

$$= \begin{vmatrix} A_y & A_z \\ B_y & B_z \end{vmatrix}\hat{i} - \begin{vmatrix} A_x & A_z \\ B_x & B_z \end{vmatrix}\hat{j} + \begin{vmatrix} A_x & A_y \\ B_x & B_y \end{vmatrix}\hat{k}$$

In this example, the second line of the 3-x-3 determinant contains the scalar coefficients of the unit vectors for vector **A** because it's the first vector listed in the cross product. The third (or bottom) line is made up of the scalar coefficients of the unit vectors in vector **B** because it's the second listed vector.

You have to assemble the contents of the determinant very carefully for this method to work. Reversing the order of **A** and **B** produces a uniquely different normal vector. That is:

$$\vec{A} \times \vec{B} \neq \vec{B} \times \vec{A}$$

Figure 6-6 is a quick illustration that I like to use to help me remember the signs on all those pesky cross products. In the three locations shown, scribble the unit vectors **i, j,** and **k.** Now, locate the two vectors you want to cross. Circle the first vector in your operation and then the second vector. The term that remains uncircled is the resulting unit vector direction of that cross product operation. Now for the cool part: If your second term is located counterclockwise from the first, the sign of the result is positive. Likewise, if the second term is located clockwise from the first term, the sign of the result is negative. If the first and second terms are both the same, the result is 0. Test it out with the equations in this section — it works!

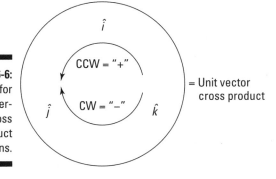

Figure 6-6:
Shortcut for remembering cross product signs.

The following equations explain the results of Figure 6-6. It shows all the combinations without requiring you to memorize the nine values below, which will hopefully help you remember them a bit more easily.

$$\hat{i} \times \hat{i} = 0 \qquad \hat{j} \times \hat{j} = 0 \qquad \hat{k} \times \hat{k} = 0$$
$$\hat{i} \times \hat{j} = \hat{k} \qquad \hat{j} \times \hat{k} = \hat{i} \qquad \hat{k} \times \hat{i} = \hat{j}$$
$$\hat{j} \times \hat{i} = -\hat{k} \qquad \hat{k} \times \hat{j} = -\hat{i} \qquad \hat{i} \times \hat{k} = -\hat{j}$$

Useful Vector Operation Identities

To further expand your toolbox for working with vectors, this section presents a few useful identities and discusses orders of operation. They all use the following vectors **A**, **B**, and **C**, such that

$$\vec{A} = A_x \hat{i} + A_y \hat{j} + A_z \hat{k}$$
$$\vec{B} = B_x \hat{i} + B_y \hat{j} + B_z \hat{k}$$
$$\vec{C} = C_x \hat{i} + C_y \hat{j} + C_z \hat{k}$$

Throughout this book, I refer to several of these properties, so you may want to dog-ear this page.

 ✔ Associative property of vector addition

$$\vec{A} + \left(\vec{B} + \vec{C} \right) = \left(\vec{A} + \vec{B} \right) + \vec{C}$$

 ✔ Commutative property of vector addition

$$\vec{A} + \vec{B} = \vec{B} + \vec{A}$$

 ✔ Distributive property of vector addition

$$a \left(\vec{A} + \vec{B} \right) = a\vec{A} + a\vec{B} \quad \text{where } a \text{ is a scalar value}$$

 ✔ Commutative property of vector dot products

$$\vec{A} \cdot \vec{B} = \vec{B} \cdot \vec{A}$$

 ✔ Anti-commutative property of vector cross products

$$\vec{A} \times \vec{B} = -\left(\vec{B} \times \vec{A} \right)$$

 ✔ Distributive property of vector cross products

$$\vec{A} \times \left(\vec{B} + \vec{C} \right) = \vec{A} \times \vec{B} + \vec{A} \times \vec{C}$$

Chapter 7

Turning Multiple Vectors into a Single Vector Resultant

*O*bjects in statics can be subjected to a wide variety of actions from an almost infinite list of sources. Having the ability to transform a system of many similar effects into a single equivalent behavior (called a *resultant*) is a truly handy skill.

Resultants aren't necessarily a cause-and-effect relationship. If you sit on a tiny, cushy kiddie chair (cause), the outcome (result) would likely be that the chair may be a bit lumpier (or in a few more pieces) than it was before.

In statics, the resultant behavior has a different meaning. Resultants represent a way of consolidating information. For example, if you sit on the kiddie chair and your friend comes in and sits on the chair at the same time, the statics resultant is that two people are applied at the same location (on your chair), or that twice the number of loads have been applied to your chair.

In this chapter, I show you several different methods for determining a statics resultant, each of which requires different techniques. Some of these methods can be labor intensive yet light on mathematical requirements, and others are more complex yet robust. However, being able to consolidate multiple vectors into one combined vector makes your calculations so much easier. Picture a statics problem as a cage full of hyper kittens — if you open the cage, the kittens run out in any number of directions and at different speeds. That's a lot of different vectors at work! Resultants give you the ability to substitute those hyperactive critters with a single replacement. In simple terms, why deal with a hundred different actions when you can deal with just one?

Getting a Handle on Resultant Vectors

A resultant vector (or simply a *resultant*) is the most basic representation of a system of combined actions that result in the same behavior as the original system. Often, this equality means you combine multiple vectors into a single equivalent vector.

Depicting a resultant vector

The first part of Figure 7-1 shows an object subjected to three separate actions in three different directions. With all of those actions combined, the body should experience a single response in a unique direction. That combined response is illustrated in the second part of the figure by V_{RES} and is oriented at its own unique angle θ_{RES}.

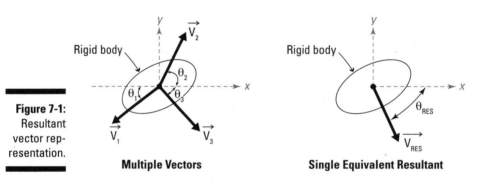

Figure 7-1:
Resultant
vector rep-
resentation.

Multiple Vectors   **Single Equivalent Resultant**

The resultant vector's *magnitude* (numerical value) and *orientation* (combined sense and direction) are uniquely determined by the magnitudes and orientations of the original uncombined vectors.

The position vector I discuss in Chapter 6 is actually a type of resultant vector. It starts at a given location and ends at a second point. However, a position vector actually has an infinite number of paths (if you ignore walls, fences, and other physical barriers) to get from one point to another. The position vector is the simplest representation of a path from one point to another, independent of which of the individual pieces you use.

Principles of resultants

The resultant calculations for a vector are the same regardless of the type of vector. Force vectors, velocity vectors, and displacement vectors can all be combined with similar types of vectors to create resultants. But before you start combining force vectors with displacement vectors to calculate resultants, keep a couple of simple guidelines in mind:

- ✔ **Only vectors of similar types may be combined to create resultants.** A resultant vector is the same vector type as the vectors that create it. Force vectors can only be combined with other force vectors — you can't combine a force vector and a velocity vector to produce a position vector resultant.

- ✔ **Magnitudes and directions of resultant vectors can be larger, smaller, or the same as any of the combined vectors.** The magnitude and direction *(sense)* of a resultant vector are directly related to the properties of the vectors that are combining to make the resultant. Two large vectors in opposite directions may produce a very minimal resultant effect.

Because a resultant is also a vector and has been created from combining smaller vectors, it has its own magnitude, sense, and point of application, which are a direct result of the vectors that helped create the resultant. For more on these properties, look at Chapter 4.

Calculating resultant magnitude and direction

You can calculate the magnitude and direction of resultant vectors in many valid ways. Generally, these techniques fall into one of three categories: graphical methods, geometric methods, and vector equation methods.

- ✔ **Graphical methods:** *Graphical methods* typically include making a detailed sketch (drawn to scale) of all the vectors in the system you want to determine the resultant for. You then make a physical measurement from the drawing to determine the solution.

- ✔ **Geometric methods:** *Geometric methods* use basic sketches and geometric principles to determine the resultant of two vectors.

- ✔ **Vector equation methods:** *Vector equation methods* use Cartesian vector notation (from Chapter 4) to determine the resultant of a system of vectors. They require the vector mathematics I explain in Chapter 6.

The most common way to illustrate a system of vector actions is the *head-to-tail* construction I first discuss in Chapter 6. This construction technique is very useful in each of the three major resultant construction techniques that I describe in the coming sections. You can perform the head-to-tail construction method by using the following steps:

1. **Select an arbitrary starting point.**

The origin of your coordinate system is often a convenient starting point, but you're not required to use that point. However, you must choose an origin somewhere.

2. **Choose any one of the original vectors and place its tail at the starting location you determined in Step 1, making sure to maintain the original orientation and sense.**

   You can start with either vector, but after you use a vector one time, you can't use it again. You must keep the same magnitude and direction for each vector.

3. **Select one of the remaining original vectors and affix the tail of that vector to the head of the vector from Step 2.**

4. **Keep attaching vectors to each other (by repeating Steps 2 and 3) until you've used all of the original vectors.**

The order of selection for the vectors that you choose in Steps 2 and 3 isn't important. You can use them in any order, but you can only use a vector once in this technique, and you must use all vectors.

Consider the system of vectors in Figure 7-2, which shows a rigid body with three vectors acting on it. There are multiple ways to find the resultant using the head-to-tail method. For illustrative purposes, I show you two different possibilities for constructing the resultant of the system of vectors. Notice that the final vector has the same magnitude, sense, and line of action for all cases.

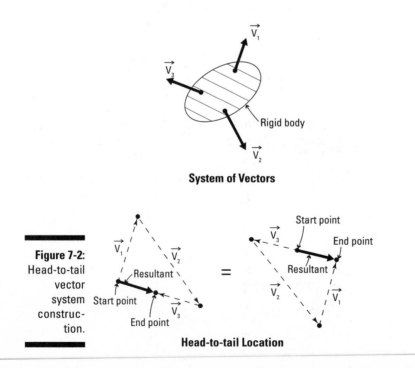

Figure 7-2: Head-to-tail vector system construction.

Using Graphical Techniques to Construct Resultants

The easiest method for determining the resultant of a system of forces is to create a scaled drawing. Graphical methods are very convenient because they typically require you to perform very few mathematical calculations and you don't need a vector formulation at all. If you can draw a line at a specified length and angle/direction, you can determine the resultant. Plus, this method can handle multiple vectors at once. Computer aided drafting (CAD) programs and hand drawings with rulers, protractors, and compasses provide useful tools for these techniques.

The graphical methods do have their drawbacks, though. They're time-consuming to draw, and you can really only use them on two-dimensional problems. Although you can theoretically use this method in three dimensions as well, it's definitely not an easy (or efficient) method for those problems. Graphical methods also require a computer or mechanical drafting tools, and you can't guarantee their numerical accuracy because your tools are only so accurate (mechanical protractors are accurate to about one-degree increments, and your magnitude measurements are always limited by the scale of your ruler). As CAD systems have become more popular, you can actually use CAD drawings to achieve more accurate numerical measurements of the magnitude and direction of the resultant. Despite some of these accuracy issues, graphical methods can provide an effective (and often quick) qualitative view of size of the magnitude and general direction of a resultant vector.

The actual procedure for construction of a graphic representation of a resultant vector is fairly straightforward. Consider the rigid body of Figure 7-3a. In this example, notice that the magnitude of each force vector is given and that the directions of those vectors are described by either angles or proportions.

To perform this construction, you utilize the head-to-tail construction technique from the preceding section with a couple of minor modifications. Just follow these steps:

1. **Choose your starting point and establish a grid for construction; indicate that your scale is one grid square represents one vector unit.**

2. **Select one of the two vectors and affix its tail to the starting location of Step 1.**

   In Figure 7-3b, I've arbitrarily chosen to start with the 1.41-Newton force. Be sure to precisely draw the force at its true scale length of 1.41 Newton at an angle of 45 degrees from the start point.

   Preserving the proper magnitude and direction angles in your sketches is very important. If your sketches aren't accurate, you can't expect your final measurements to be accurate!

WARNING!

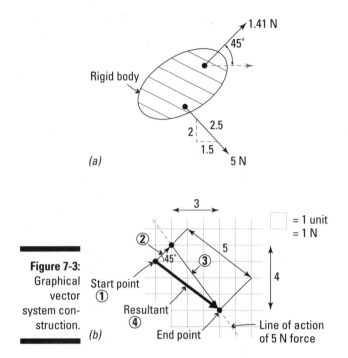

Figure 7-3:
Graphical
vector
system con-
struction.

3. **Sketch the next vector to scale at the proper orientation; repeat for any additional vectors.**

Start by laying in the *line of action* (line in space where the vector is acting) of the force. You can easily find this line in the example in Figure 7-3b because you actually know the proportions of the grid.

To establish this line that passes through the end point of the vector from Step 2, follow these steps:

A. Locate a second point by counting down the number of grid squares on the vertical leg of the *proportion triangle* (which I discuss in Chapter 5) and then counting over the number of grid squares that are on the horizontal leg of the proportion triangle.

B. Draw a continuous line of action through these two defined points to establish the direction of the vector you're working with and then measure a distance that has the same length as the magnitude of this vector along the line of action from the end point of the vector in Step 2.

In the example, you measure five units from the end point of the 1.41-Newton force. This new point is the head location for the 5-Newton force.

C. Draw your vector arrow from the tail point to the new head location.

Remember that proportion triangles tell you the direction of the line of action of the force. You can then use the magnitude to determine how far down that line of action the head is from the tail of the force vector.

4. **Sketch the resultant by drawing a vector arrow from the start point of Step 1 to the end point of the last vector that you drew.**

Congratulations — you've determined the direction of the resultant vector! You can now use a protractor to measure the angle of the resultant vector relative to a horizontal gridline, a vertical gridline, or even one of the original vectors' lines of action. The choice is yours. To determine the magnitude of the resultant, you simply need to use a ruler to measure the length of the resultant vector's arrow on your grid.

Using Geometric Methods to Construct Resultants: The Parallelogram Method

To avoid the problems inherent in graphical solutions (see the preceding section), geometric methods can present a better estimate of the magnitude and direction of a resultant. They don't require vector formulations, and the solution method is very, well, methodical. Unfortunately, these techniques only work effectively on two-dimensional problems and only on a maximum of two vectors at a time (although you can repeat the process with your resultant and additional vectors until only one resultant remains).

The most popular geometric construction technique is dubbed the parallelogram method because it incorporates an actual parallelogram into the solution process. However, you need to master several trigonometric identities to use this technique, so I tackle those in the following sections before showing you how to create the resultant.

Useful geometric relationships

By far the most essential geometric identities you need, aside from SOHCAHTOA (which I explain in Chapter 2), are the law of cosines and the law of sines. Each of these methods utilizes angle and dimensional relationships for triangles such as the one shown in Figure 7-4, which shows a triangle *123,* having sides of length *A, B,* and *C* and corresponding opposite angles of *a, b,* and *c,* respectively.

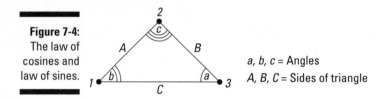

Figure 7-4:
The law of
cosines and
law of sines.

a, b, c = Angles
A, B, C = Sides of triangle

Law of cosines

The law of cosines provides a formula that relates two sides and an *included angle* (an angle between two adjacent sides of a polygon) to the length of the third side. These formulas can be expressed as

$$A^2 = B^2 + C^2 - 2BC\cos(a)$$
$$B^2 = A^2 + C^2 - 2AC\cos(b)$$
$$C^2 = A^2 + B^2 - 2AB\cos(c)$$

Notice in Figure 7-4 that each of these expressions includes the length of two sides, and the angle between those same two sides. From this information you can then determine the length of the third side.

You must know at least two sides and the angle between those two sides to compute the third side by using the law of cosines. However, if you happen to know all three sides of the triangle but none of the angles, you can actually solve for the included angle by rearranging the equation and using an inverse cosine function (or cos⁻¹). For example, to find angle a

$$a = \cos^{-1}\left(\frac{B^2 + C^2 - A^2}{2BC}\right)$$

Similarly, you can create expressions for the angles b and c by rearranging their respective law of cosines equations.

Is that the Pythagorean theorem I see?

An interesting fact pops up when you apply the law of cosines to a *right triangle* (which has a 90-degree angle). If you substitute c = 90 degrees into the last equation, you get the following:

$C^2 = A^2 + B^2 - 2AB\cos(90) = A^2 + B^2$ (because trigonometry tells you that $\cos(90) = 0$), or $C^2 = A^2 + B^2$, which is a classic representation of the Pythagorean theorem for right triangles. For more information on the Pythagorean theorem, you can look at Chapter 2.

Law of sines

The law of sines is another extremely useful geometric relationship, given by the relationships

$$\frac{\sin(a)}{A} = \frac{\sin(b)}{B} = \frac{\sin(c)}{C}$$

where the parameters of this expression are the same parameters shown in Figure 7-4. This relationship basically says that the ratios of a side of a triangle with respect to the sine of its *opposite angle* (the angle directly across from that side) are always constant for any given triangle. To use the law of sines, you must know at least one side of a triangle and its corresponding opposite angle to establish the ratio. Then, to actually solve for unknown parameters, you simply need to know either another side or another angle of the triangle. You can then rearrange the expression and solve for one of the unknowns; I show you how to do that in more detail in the following section.

The parallelogram method

The parallelogram method is a basic geometric method for determining the resultant of two vectors by constructing a parallelogram. You locate the resultant force and then use the laws of sines and cosines to perform calculations on one half (a triangular section) of the parallelogram. A *parallelogram* is a specific variation of the *quadrilateral* (four-sided geometric shape) and is shown in Figure 7-5.

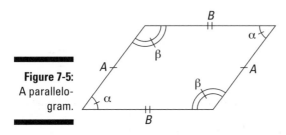

Figure 7-5:
A parallelogram.

A parallelogram by definition has the following properties:

- ✔ Opposite sides of the parallelogram are *congruent* (or equal) and must be parallel.

- ✔ Opposite angles of the parallelogram are also congruent.

- ✔ The total sum of the internal angles of a parallelogram must add up to 360 degrees.

Figure 7-6 shows a block subjected to a force of 200 Newton at an angle of 45 degrees from the vertical, and a second force of 350 Newton at an angle of 30 degrees below the horizontal

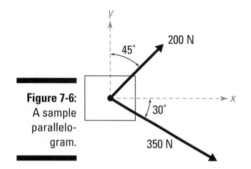

Figure 7-6:
A sample
parallelo-
gram.

To use the parallelogram method, you follow these basic steps:

1. **Connect the two known force vectors at their tail locations, preserving their orientation and magnitude.**

   You connect the two vectors at their tails and place this point at the origin of your Cartesian axes as shown in Figure 7-7.

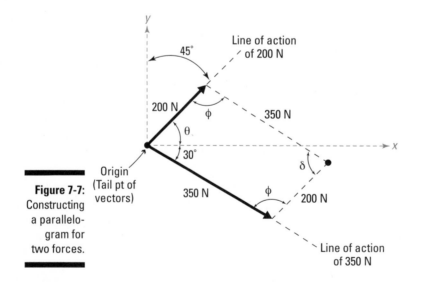

Figure 7-7:
Constructing
a parallelo-
gram for
two forces.

2. **Construct a parallelogram where one side is the magnitude of one of the vectors in its current position and the other side is the magnitude of the other vector in its current position.**

   In this example, one side of the parallelogram is 200 Newton and the other is 350 Newton, as shown in Figure 7-7. You may have to pick up one of the forces and move it such that the tails of the vector arrows are acting at the same location. You can do that because forces are *sliding vectors* (or vectors that can act anywhere along their lines of action), so you can actually slide them along their respective lines of action until their tails meet at a common point.

3. **Compute the angle δ between the vectors.**

   The angle θ is the difference of 90 degrees between the *x*- and *y*-Cartesian axes and the 45-degree orientation of the 200-Newton force (so θ = 90° − 45° = 45°). The angle δ is the angle between the 200-Newton and 350-Newton forces and is the sum of θ + 30 degrees, or 75 degrees. Remember that the 30 degrees is the orientation angle of the 350-Newton force below the *x*-axis.

4. **Compute the angle φ for the remaining angles.**

   Because you know all quadrilaterals must have a total of 360 degrees in interior angles, you can compute the remaining unknown interior angle as follows:

   $$\delta + \delta + \phi + \phi = 360°$$

   $$75° + 75° + 2\phi = 360°$$

   $$2\phi = 210°$$

   $$\phi = 105°$$

5. **Draw the resultant force vector as the diagonal of the parallelogram.**

   Because the parallelogram contains two diagonals, you select the one that shares the tail points of the original vector. In this example, the tail of the resultant is at the same point as the tail of the original 200-Newton and 350-Newton vectors, and the head of the resultant is at the opposite corner, as you can see in Figure 7-8).

6. **Remove one of the triangular regions from the parallelogram that was created when you sketched the resultant force vector in Step 5.**

   Which triangle you select makes no difference because they're both the same. The top triangle has the same angles and sides as the bottom one. Include any dimensional or angular information that you may know or have previously calculated. In Figure 7-9, I've chosen the upper triangle for my calculations in this example.

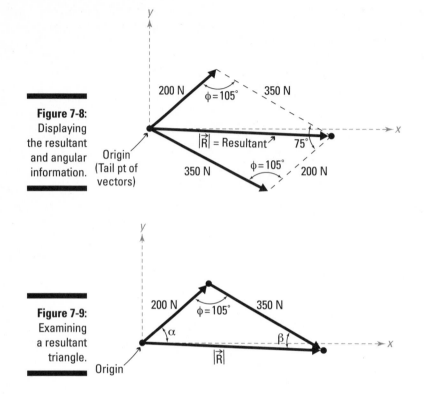

Figure 7-8:
Displaying
the resultant
and angular
information.

Figure 7-9:
Examining
a resultant
triangle.

7. **Use the law of cosines to determine the magnitude of the resultant force.**

   After splitting the parallelogram into a force triangle consisting of the resultant and two sides, you know at least one of the angles of the triangle and the two adjacent sides from the original 200-Newton and 350-Newton vectors. In Figure 7-9, you know two sides of the triangle and the included angle of 105 degrees. You can plug this info into the law of cosines; your result looks something like the following:

$$\left|\vec{R}\right| = \sqrt{(200 \text{ N})^2 + (350 \text{ N})^2 - 2(200 \text{ N})(350 \text{ N})\cos(105°)} = 445.81 \text{ N}$$

8. **Use the law of sines to compute the remaining angle α to help establish the direction of the resultant.**

   This angle gives the orientation of the resultant relative to the original 200-Newton force vector. At this point, you now know all three side dimensions and one of the related angles. Using the law of sines, you can establish the triangle's ratios.

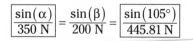

The boxes in this equation just help you remember the pieces you need to solve for. Because you know the 105-degree angle and the opposite side of 445.81 Newton, you can definitely use those two values to establish the ratio. Because you want to compute the angle α, you must use the ratio term that includes that parameter. A bit of rearranging produces the calculation for the desired angle:

$$\sin(\alpha) = \left(\frac{\sin(105°)}{445.81 \text{ N}} \right) \cdot (350 \text{ N})$$

$$\alpha = 49.31°$$

9. **Define your reference line for the angle measurements.**

   Measuring from the x- or y-axis is usually the most common reference location. However, you can also measure the direction angle of a resultant relative to the angle of any other vector.

   If you want to base the calculation from the vertical Cartesian y-axis, you can compute the measured angle from

   $$\theta_y = 45° + 49.31° = 94.31°$$

   which is measured clockwise from the positive Cartesian y-axis. Similarly, if you want to base your reference from the positive Cartesian x-axis, you can report the reference angles as

   $$\theta_x = 90° - \theta_y = 90° - (45° + 49.31°) = -4.31°$$

   In this case, the angle is a negative value, which indicates that it occurs below the positive Cartesian x-axis. Either direction is a correct representation, and both are displayed in Figure 7-10.

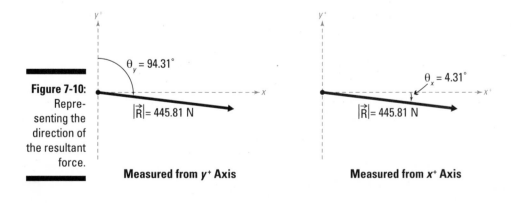

Figure 7-10: Representing the direction of the resultant force.

$\theta_y = 94.31°$

$|\vec{R}| = 445.81$ N

Measured from y^+ Axis

$\theta_x = 4.31°$

$|\vec{R}| = 445.81$ N

Measured from x^+ Axis

Define your reference clearly. Be sure to include a sketch that properly illustrates where you're measuring from, and in which direction.

Using Vector Methods to Compute Resultants

One of the most consistent ways of determining a resultant force is by using the vector addition I explore in Chapter 6. You can use this technique with as many vectors as your heart desires, and it works with two- and three-dimensional problems (unlike the other two techniques in this chapter, which only work easily in two dimensions). It requires only basic math skills — if you can add, you can do vector addition — and it doesn't generally have the accuracy concerns of the graphical techniques (see "Using Graphical Techniques to Construct Resultants" earlier in the chapter). *Note:* You may need to use some basic trigonometry to create the original vectors in the first place (as I show in Chapter 5), but after that's done, actually computing the resultant is a breeze!

The greatest advantage of the vector methods is that you don't have to find perpendicular distances or worry about maintaining accurate information about the sense of the vector as you do with the other methods in this chapter. All that information is built into the notation of the vector itself. In fact, if you can create vectors, the hardest part of your statics work is already complete because all you need to do is use vector addition to compute the combined result. As long as your calculator has batteries and you can create the necessary Cartesian vectors, finding the resultant vector should be a snap!

Consider the system of two vectors **A** and **B** in Figure 7-11, which shows a simple vector addition case. Vectors **A** and **B** are combined to create a resultant vector **C**. However, with vector addition, you can also find the resultant of multiple vectors all at once. Utilizing the same head-to-tail construction technique I discuss in the graphical methods section earlier in this chapter, you can quickly determine the resultant for the system. However, if you know the vector representation for each of the forces **A, B, C,** and **D,** you can find the resultant as being from the tail of vector **A** to the head of vector **D.**

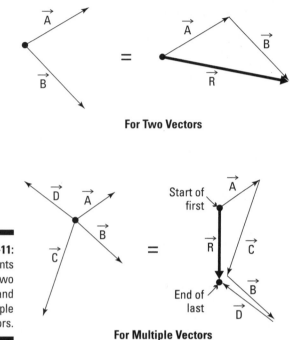

For Two Vectors

Figure 7-11: Resultants of two vectors and of multiple vectors.

For Multiple Vectors

Using vector addition

If you have the vectors defined in vector notation with all the **i, j,** and **k** pieces already determined (see Chapter 5), you can calculate resultants with the following vector addition formula:

$$\vec{R} = \vec{A} + \vec{B}$$

Even better, you can expand this simple example to easily handle multiple vectors in exactly the same way:

$$\vec{R} = \vec{A} + \vec{B} + \vec{C} + \vec{D} + \ldots = \Sigma(\text{vectors})$$

Consider the example of Figure 7-12, which shows a particle subjected to three forces $\mathbf{F_1}$, $\mathbf{F_2}$, and $\mathbf{F_3}$.

To find the resultant of these vectors, you simply add them together:

$$\vec{R} = \vec{F_1} + \vec{F_2} + \vec{F_3} = \left(10\hat{i} + 10\hat{j} + 10\hat{k}\right) \text{lbs} + \left(5\hat{i} - 6\hat{j} + 8\hat{k}\right) \text{lbs}$$
$$+ \left(-10\hat{i} + 2\hat{j}\right) \text{lbs}$$

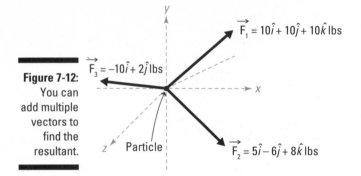

Figure 7-12:
You can
add multiple
vectors to
find the
resultant.

Gathering all of the terms in front of **i, j,** and **k,** you can then simplify the
expression to

$$\vec{R} = (10 + 5 - 10)\hat{i} + (10 - 6 + 2)\hat{j} + (10 + 8 + 0)\hat{k} \text{ lbs}$$
$$= 5\hat{i} + 6\hat{j} + 18\hat{k} \text{ lbs}$$

Thus, the resultant of the three vectors of this example is **5i** + **6j** + **18k,** which
itself is in vector notation. From here, you can then find the magnitude of the
vector by using the following equation:

$$\left| \vec{R} \right| = \sqrt{(5)^2 + (6)^2 + (18)^2} = 19.62 \text{ lbs}$$

Calculating the direction of the vector resultant

After you've computed the vector form and the magnitude of the resultant,
you can define the direction in any of the ways I define in Chapter 5. Perhaps
the easiest is to simply create a unit vector from this information. For the
example in the preceding section, you need to divide the resultant vector by
its magnitude to create the unit vector that describes the direction. You can
then compute the direction cosines to determine the angles if you so desire.
The following equation shows you how to create a unit vector for Figure 7-12
in the previous section.

$$\overrightarrow{u_R} = \frac{5\hat{i} + 6\hat{j} + 18\hat{k}}{19.62} = 0.255\hat{i} + 0.306\hat{j} + 0.917\hat{k}$$

Chapter 8

Breaking Down a Vector into Components

*I*n Chapter 7, I show you how to take multiple vectors and combine them into a single resultant behavior, which is a useful skill in helping simplify the number of actions on an object. That works fine if you're interested in examining the combined behaviors of an object, but what happens when you're interested in studying multiple behaviors but only have a single resultant to work with? For this situation, you need to understand how to create multiple behaviors of a resultant, or the components of a resultant vector.

The most useful feature of working with components is that these behaviors let you explore basic behaviors in more detail. For example, when an airplane is coming in for a landing, its approach is actually a vector with a given orientation at a specific speed. However, the pilot must maintain a certain horizontal behavior (which ensures that the plane actually reaches the runway and doesn't overshoot) to land the plane safely while guaranteeing that the vertical descent isn't so fast that it causes the plane to crash into a fiery heap when it hits the ground. The pilot needs to be aware of both the vertical and horizontal behaviors at the same time, for uniquely different reasons.

In this chapter, I show you how to break a single vector back into multiple behaviors that act entirely in Cartesian or non-Cartesian directions.

Defining a Vector Component

You may be asking yourself, "Why on earth would I care about breaking one combined action into several smaller actions?" Answer: In many statics and physics problems, you're often interested in examining these individual actions on a case-by-case basis. These individual actions of a single resultant vector are known as *components.* For example, consider the projectile motion example shown in Figure 8-1.

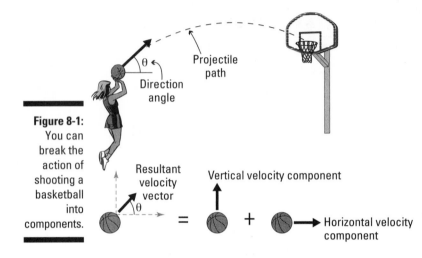

Figure 8-1: You can break the action of shooting a basketball into components.

In this example, a player is shooting a basketball at a hoop in the distance. When she attempts the shot, the ball has a unique velocity *(magnitude)* and direction angle θ, which defines the initial path of the projectile *(sense and line of action)* at the exact instance she releases the ball. Together, these three pieces of information define the initial velocity vector of the basketball. (Flip to Chapter 4 for more on these three main vector properties.)

Because the player remains stationary as she attempts the shot (this isn't a slam-dunk attempt), the problem you want to examine is what's happening to the basketball itself — especially whether the path results in a scoring shot.

From experience, you should be able to recognize that as the player begins the shot, the ball moves upward with a certain vertical velocity while moving away from her with another horizontal velocity. These two velocities are the horizontal and vertical components of the combined (or *resultant*) velocity vector for the basketball.

In two dimensions, you need to represent two components for every resultant vector, and in three dimensions you actually have three components that you need to determine. However, for both two- and three-dimensional cases, these components together must always result in the same original combined behavior.

A component of a resultant vector must also be the same type of vector as the resultant itself. If you're finding the components of a force vector, the components are also force vectors.

In the parallelogram method discussion in Chapter 7, I show you how to create a resultant vector of two vectors by constructing a basic parallelogram where the resultant is the diagonal across the parallelogram that shares the tail point of the original two vectors. The two sides of the parallelogram (the original vectors) that share the tail of the resultant are actually components of that resultant vector, as shown in Figure 8-2.

Figure 8-2: Components of a resultant and parallelogram.

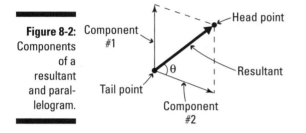

Resolving a Vector into Cartesian and Non-Cartesian Components

The process of creating a vector component is known as *resolving* a vector. In working with vectors in statics, you always resolve two-dimensional actions such as forces and displacements into exactly two pieces for two-dimensional problems and into exactly three pieces for three-dimensional problems. (You can also depict two-dimensional actions with three-dimensional components by making a third component that has zero magnitude.)

To accomplish this task, you need to decide which type of component is actually required. These types of components are often separated into two categories: Cartesian (or rectangular) components and non-Cartesian components.

> ✔ **Cartesian components:** As the name *Cartesian components* implies, all of the resolved vector components are aligned with the Cartesian *x*-, *y*-, and (for three-dimensional problems) *z*-axes of your coordinate system. Sometimes you see them referred to as *rectangular components*. Rectangular components are probably the most common components you calculate and, fortunately, are usually the easiest to compute.

> ✔ **Non-Cartesian components:** *Non-Cartesian components* of a vector aren't necessarily aligned with the Cartesian-axes. One or more components may be aligned with the Cartesian axes, but at least one is not. (If all were aligned with the axes, they'd be Cartesian components.)

In later chapters, I help you with the actual selection process for choosing the type of components to help you best solve a particular problem, and I give you some additional pointers on choosing the appropriate directions of your components. But for now, in the following sections I focus on explaining the specific calculation techniques you actually employ after you determine the type of component.

After you compute the vector components, you can begin to compute their magnitudes, and represent them with vector equations.

Using Cartesian concepts to calculate Cartesian components

To determine the components in two dimensions, you need to follow a few simple steps to create right triangles:

1. **Form a right triangle from the vector by using a line parallel to the *x*-axis attached at the tail and a second line parallel to the *y*-axis and attached to the head of the vector.**

2. **Draw the *x*-component vector by drawing from the tail point of the original vector.**

   Locate the head of the *x*-component vector at the intersection of the horizontal and vertical lines.

3. **Draw the *y*-component vector by drawing from the intersection of the horizontal and vertical lines (the tail point).**

   Place the head of the *y*-component vector at the head point of the original vector.

Figure 8-3 shows the finished diagram.

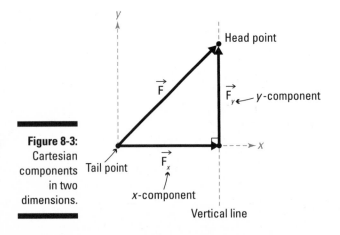

Figure 8-3: Cartesian components in two dimensions.

To determine the components in three dimensions, you use a similar head-to-tail construction technique.

1. **Form a cube around the vector by placing one corner at the vector's tail and one at the opposite diagonal corner at the vector's head.**

2. **Establish reference lines by drawing a vertical line through the head point of the original vector; at the point where this vertical line crosses the horizontal plane, draw one horizontal line parallel to the *x*-axis and a second horizontal line parallel to the *z*-axis.**

   In the example in Figure 8-4, your first reference line is parallel to the *y*-axis and crosses the horizontal plane *xz*.

3. **Determine the *x*-direction vector components by drawing the *x*-component vector from the tail point of the original vector.**

   In Figure 8-4, you place the point labeled Int. #1 at the head of the *x*-component, which represents the location where a horizontal line drawn parallel to the *z*-axis intersects with the *x*-axis of the reference coordinate system.

4. **Establish the *z*-direction vector component by drawing the *z*-component vector from the head of the *x*-component to the point where the two horizontal lines intersect.**

   In this example, you draw from Int. #1 to Int. #2.

5. **Determine the *y*-direction vector component by placing its tail at the horizontal intersection and its head at the head point of the original vector.**

   In Figure 8-4, you place the tail of the *y*-component vector at Int. #2.

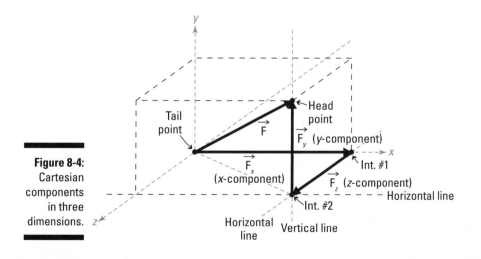

Figure 8-4:
Cartesian components in three dimensions.

Figuring component magnitudes

For two-dimensional problems, the process of resolving a vector into components is as simple as drawing a right triangle. Consider the example in Figure 8-5a, which shows a force of 250 Newton acting at an angle of 130 degrees from the positive x-axis of the Cartesian coordinate system, which is the same as measuring 50 degrees from the negative x-axis.

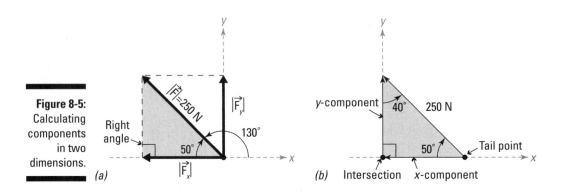

Figure 8-5:
Calculating
components
in two
dimensions.

To resolve this force vector into its rectangular components, you need to first locate a right triangle, indicated by the shaded region in Figure 8-5a. The original vector is located on the hypotenuse of this right triangle, and the component vectors are then drawn along the other two edges of the right triangle as shown in Figure 8-5b.

Using the principles of SOHCAHTOA (refer to Chapter 2), you can calculate the magnitude of the x-component of this from

$$\left|\vec{F_x}\right| = (250 \text{ N}) \cdot \cos(50°) = 160.69 \text{ N}$$

where the angle 50 degrees is measured from the negative x-axis.

Similarly, you can compute the y-component,

$$\left|\vec{F_y}\right| = (250 \text{ N}) \cdot \sin(50°) = 191.51 \text{ N}$$

The values that you just computed are only the scalar magnitudes of the components and don't include the sense or direction of the component. For two-dimensional problems, you need to create component vectors or assign the directions by using simple logic and the Cartesian unit vectors **i, j,** and **k,** as I discuss in Chapter 5.

Using the bracket notation method of three-dimensional vector representation

Statics provides a few ways to represent vector information, including **i**, **j**, and **k** unit vector notations and scalar magnitudes and associated directions I discuss in Chapter 5.

But you'll see another bracket notation method pop up from time to time in calculus and physics textbooks, and it can be useful at times in statics. The major advantage of this notation is that the scalar components in each of the principle directions are readily visible and require no additional calculations to determine them.

For example, consider the acceleration vector **a** written in Cartesian notation:

$$\vec{a} = +2.5\hat{i} - 3.1\hat{j} + 6.2\hat{k} \text{ m/s}^2$$

The bracket representation for this vector is:

$$\vec{a} = \langle +2.5, -3.1, +6.2 \rangle \text{ m/s}^2$$

If you compare the Cartesian notation with the bracket notation, you see that each term in the bracket is actually one of the three scalar components of the Cartesian vector notation. In this acceleration example, the first term in the brackets, or the *x*-component, is +2.5 meters per second squared; the second term, or the *y*-component, is −3.1 meters per second squared; and the third term, or the *z*-component, is +6.2 meters per second squared.

Using scalar magnitudes and directions to create vector components

In Figure 8-5, the *x*-component of the force has a magnitude of 160.69 Newton. However, looking at the vector representation of the *x*-component, you can clearly see that the component is acting to the left, or in the negative *x*-direction. Because you can denote the negative *x*-direction by using the −**i** unit vector, you can create the *x*-component force vector:

$$\vec{F_x} = \left|\vec{F_x}\right| \cdot \vec{u_x} = 160.69\text{N} \cdot \left(-\hat{i}\right) = -160.69\hat{i} \ N$$

You calculate the *y*-component vector in a similar fashion. Recognizing that the *y*-component of this vector is acting in a positive *y*-direction, you can use a positive **j** unit vector to describe the direction.

$$\vec{F_y} = \left|\vec{F_y}\right| \cdot \vec{u_y} = 191.51 \text{ N} \cdot \left(+\hat{j}\right) = +191.51\hat{j} \ N$$

Computing vector components in three dimensions

When you resolve a three-dimensional vector into its rectangular components, the components must be mutually perpendicular to each other, which usually means each component is parallel to one of the three Cartesian axes.

Determining the components of a three-dimensional vector is actually fairly simple as well. For example, if you wanted to find the rectangular or Cartesian components of the following velocity vector:

$$\vec{V} = -2.50\hat{i} + 3.60\hat{j} + 5.10\hat{k} \text{ m/s}$$

you simply need to strip off the numerical values that occur before the **i, j,** and **k** values in the vector representation.

$$\left|\vec{V_x}\right| = 2.50 \text{ m/s acting in the } -\hat{i} \text{ direction}$$

$$\left|\vec{V_y}\right| = 3.60 \text{ m/s acting in the } +\hat{j} \text{ direction}$$

$$\left|\vec{V_z}\right| = 5.10 \text{ m/s acting in the } +\hat{k} \text{ direction}$$

Determining components on a non-Cartesian orientation

Although Cartesian vectors always work with any statics problem you encounter, sometimes they aren't the most efficient tool in your proverbial toolbox.

The major advantage of using non-Cartesian components is that it allows you to use a more convenient coordinate system that may better match the symmetry of the object. For example, consider the force from a shockwave caused by an explosion. In this type of phenomenon, the shockwave moves in three dimensions, radially, away from the center of the explosion. If you use Cartesian representation to portray this force, you need three vector components for every small area on the surface of the shockwave, and each point on the surface has a different set of three components. However, if you chose to work in a different, non-Cartesian coordinate system such as spherical coordinates, you can transform this complex three-dimensional problem into a more simplified one-dimensional situation.

Calculating non-Cartesian components of two-dimensional vectors

In this section, I show you how to calculate a non-Cartesian component and its corresponding Cartesian component. Figure 8-6 shows the same 250-Newton force from Figure 8-5 earlier in the chapter oriented at an angle of 130 degrees from the positive Cartesian *x*-axis (or shown as 50 degrees from the negative *x*-axis).

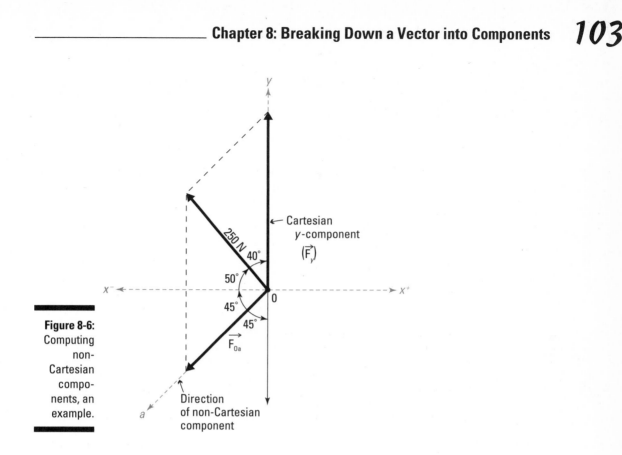

Figure 8-6:
Computing
non-
Cartesian
compo-
nents, an
example.

In this example, I've arbitrarily chosen to find one component in the direction of the Cartesian *y*-axis and the other, non-Cartesian component along the line Oa that is oriented 45 degrees below the negative *x*-axis. The following sections show you how you can use some of the resultant techniques from Chapter 7 to determine the magnitudes of these components.

Using the parallelogram method

In the parallelogram method, you basically construct a parallelogram with sides in the direction of the *y*-component and the component in the direction of Oa. Referring to Figure 8-6, you already know the resultant (250 Newton) and its direction (50 degrees from the horizontal). From geometry, you can conclude that the angle between $\mathbf{F}_y$ and $\mathbf{F}_{Oa}$ is 45 degrees as shown in Figure 8-7a. You can then pull the force triangle from the parallelogram and determine the angles geometrically as shown in Figure 8-7b.

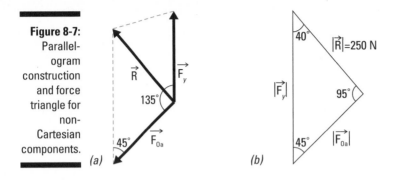

Figure 8-7:
Parallel-
ogram
construction
and force
triangle for
non-
Cartesian
components.

(a) (b)

With this information, you can find the components $\mathbf{F_y}$ and $\mathbf{F_{Oa}}$ from the law of sines.

$$\frac{\left|\vec{F_y}\right|}{\sin(95°)} = \frac{\left|\vec{F_{Oa}}\right|}{\sin(40°)} = \frac{250 \text{ N}}{\sin(45°)}$$

From this relationship, you can then compute the scalar components:

$$\left|\vec{F_{Oa}}\right| = 227.26 \text{ N}$$

$$\left|\vec{F_y}\right| = 352.21 \text{ N}$$

Using Cartesian techniques to find non-Cartesian components

An easier technique involves breaking each of these vectors into their respective x- and y- Cartesian components. First, you need to convert the resultant vector **R** into Cartesian form as shown in the following equation and Figure 8-8:

$$\vec{R} = -250\cos(50°)\hat{i} + 250\sin(50°)\hat{j} \text{ N} = -160.70\hat{i} + 191.51\hat{j} \text{ N}$$

Even though you don't know the magnitudes of the components, you can still create a Cartesian vector form. Just keep the magnitudes in the equations as a variable:

$$\vec{F_{Oa}} = -\left|\vec{F_{Oa}}\right|\sin(45°)\hat{i} - \left|\vec{F_{Oa}}\right|\cos(45°)\hat{j}$$

$$\vec{F_y} = 0\hat{i} + \left|\vec{F_y}\right|\hat{j}$$

Remember, you can find the resultant vector by performing vector addition (see Chapter 6):

$$\vec{R} = \vec{F_{Oa}} + \vec{F_y}$$

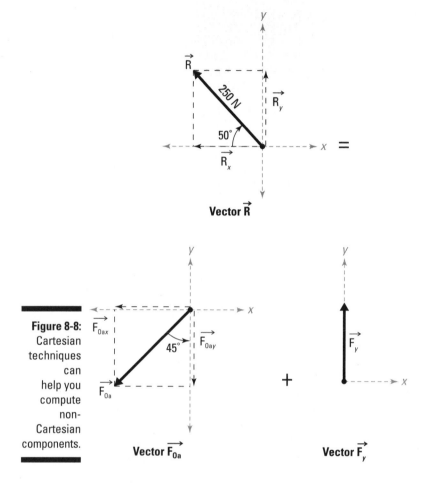

Figure 8-8: Cartesian techniques can help you compute non-Cartesian components.

You can then compare all the terms in front of the **i** unit vector, and all of the terms in front of the **j** unit vector to create a system of two equations that you can solve simultaneously.

$$-160.70\ \text{N} = -\left|\overrightarrow{F_{Oa}}\right|\sin\left(45°\right) + 0$$

and

$$191.51\ \text{N} = -\left|\overrightarrow{F_{Oa}}\right|\cos\left(45°\right) + \left|\overrightarrow{F_y}\right|$$

Solving for the two unknown magnitudes gives you $\left|F_{Oa}\right| = 227.26$ N and $\left|F_y\right| = 352.21$ N, which are the same results as the parallelogram method example in the preceding section.

Components always come in pairs, and their magnitudes are directly related to each other. If you resolve a two-dimensional resultant into components, you *must* include both components in your calculations.

In this example, you calculated a Cartesian *y*-component and found it to be 352.21 Newton when paired with the non-Cartesian component along the line Oa. However, when you calculate the components for the same original force vector (in the earlier section "Using Cartesian concepts to calculate Cartesian components") the *y*-component only had a magnitude of 191.51 Newton for the same original force vector as this example. By comparison, these two *y*-components are significantly different. Depending on the direction of their counterpart components, two components acting in the same direction can vary quite dramatically. For this reason, you must always work with all of the components for a single force at the same time.

Part III
Forces and Moments as Vectors

The 5th Wave   By Rich Tennant

"With any luck the rope will break and we'll see a perfect example of force and moment acting on an object – in this case, a freshman."

In this part . . .

The chapters in this part introduce you to forces and moments, which are common effects that influence rigid bodies. I explain the different types of translational effects, known as *forces*, and how to compute the *resultant* (that is, the combined effect) of a distributed force; to further your resultant prowess, I also demonstrate how to determine the location of this resultant through centroid (or center of area) calculations. In addition, I show you multiple ways of computing a rotational effect known as a *moment,* including using vectors and utilizing scalar values.

Chapter 9

Applying Concentrated Forces and External Point Loads

*A*s you move forward with your modeling of the world around you, which is a major step in the application of statics, one of the biggest obstacles you need to overcome is determining how to apply those pesky forces to your objects. Forces come from a wide range of sources and occur both internally and externally on the object, so you want to understand several basic rules for how these forces are applied.

In this chapter, I explain the differences between internal and external forces, explain more about where these forces come from, and show you several common scenarios in which you encounter them. I also show you some simple steps for working with forces in ropes, cables, and springs, as well as how to handle gravitational forces on objects (or self weight). Finally, I introduce the principle of transmissibility, which allows you to move forces to different locations on an object and proves very useful in statics. This chapter serves as a stepping-stone into crafting the pictures you use when applying statics to the world around you (which I cover in Part IV).

Comparing Internal and External Forces and Rigid and Deformable Bodies

You can typically separate forces into two basic categories: internal and external.

✔ **External forces:** *External forces* are created by an action external to the object itself. Examples of external forces include the force created by a ball striking a wall or the weight of snow resting on the roof of your house. External forces are often the easiest to calculate because they're usually caused by a visible phenomenon and can be measured.

✔ **Internal forces:** *Internal forces* are developed inside the object as a response to the applied external loads (or external forces) and system restraints. Tension in a cable, compression in a column, and torque in a drive shaft are all examples of internal forces. One of the major difficulties in working with internal forces is determining the type and direction of each; I show you some of the techniques for identifying internal forces beginning in Chapter 20.

Because internal and external forces are both force vectors, you need to include all the usual information such as *magnitude* (the numerical value of the vector), *sense* (direction of the vector's action), and *point of application* (physical location where the vector is acting), as I discuss in Chapter 4. In addition to distinguishing between internal and external forces, you have to make an assumption about the type of system you're dealing with. *Rigid bodies* are objects that don't experience deformation when the vector acts upon them. If internal forces act on a *deformable body,* however, the object undergoes a physical change or deformation, which can alter both your internal and external load calculations or change any relevant geometric dimensions.

In reality, all objects are deformable, but for the simple purpose of understanding the basic fundamentals of statics analysis, here I assume my objects are always rigid bodies.

Exploring External Concentrated Forces

A *concentrated force* is a force that acts at a point, or more realistically, on a very small area. Drawing an external concentrated load requires a single arrowhead, as shown in Figure 9-1. Remember that you usually place the vector's tail at the point of application on the object.

In Figure 9-1a, a man is pulling on a rope with a tension of 100 Newton; this force causes the rope to tighten. In this case, you place the single-headed arrow such that the tail of the force is applied where the man's hands attach to the rope.

Conversely, in Figure 9-1b, a 250-pound man is sitting on a chair. Because the force of the man is pushing down on the chair, the single-headed arrow's head is applied at the point of application.

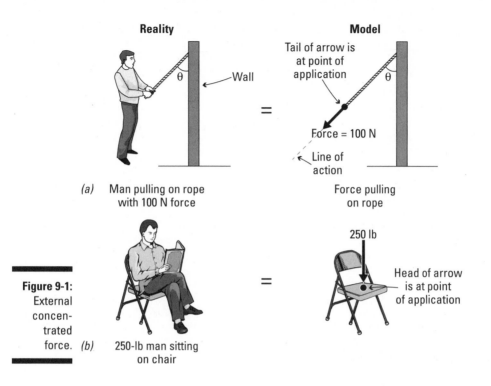

Reality

Model

Tail of arrow is at point of application

Wall

θ

θ

=

Force = 100 N

Line of action

(a) Man pulling on rope with 100 N force

Force pulling on rope

250 lb

Head of arrow is at point of application

=

Figure 9-1:
External concentrated force.

(b) 250-lb man sitting on chair

In reality, your calculations produce the same equations and solutions whether the head or the tail of the vector is applied at the point of application as long as the sense, magnitude, and line of action are all the same.

The following sections introduce several types of concentrated forces.

Normal forces from contact

In statics, you idealize most contact forces as being concentrated forces. *Normal contact forces* are forces that develop from one object pushing on another. They can be caused by self weight, applied external load, or any other force. To illustrate this idea, consider the block sitting on the ground in Figure 9-2a.

To see the contact forces, you need to separate the block from the ground in the picture. In Figure 9-2b, the force from the block pushes down on the ground. Conversely, if you examine the block by itself (as in Figure 9-2c), you see that in order to keep the block from falling through the ground, a second contact force is required to hold the block in its position. This reaction force from the ground onto the block is balanced by the contact force from the block onto the ground.

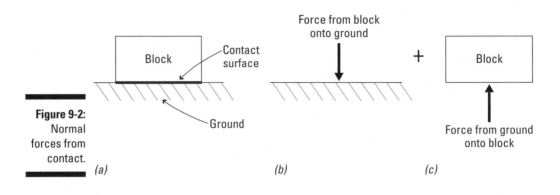

Figure 9-2:
Normal
forces from
contact.

(a)   *(b)*   *(c)*

Contact forces for both the ground and the block are always perpendicular to the contact surface. For this reason, you sometimes see contact forces referred to as *normal forces* because they're normal or perpendicular to the surface.

The point of application of contact forces is always somewhere along the contact surface, but it's exact location depends on the type of problem — you want to pay special attention to this detail when you encounter friction problems (such as the ones I cover in Chapter 24) and rotation problems (such as those in Chapter 12).

Friction

Friction is a type of external force that develops from resistance as one object tries to slide past another. From experience, you know that if you go up to an object such as a refrigerator and push on it with a very small force, it doesn't move. However, if you continue to push and eventually push hard enough, the refrigerator starts to slide. It's those hidden friction forces that initially prevent the refrigerator from sliding. For a visual of this concept, consider the block in Figure 9-3a, which is subjected to a force that causes it to begin sliding along the ground.

Just like for normal contact forces, you can separate the block from the ground and expose the internal forces. The sense of the friction force on the object that wants to move goes against the direction of the motion of the object (as shown in Figure 9-3b). The friction force on the opposite side of the contact surface (on the ground in Figure 9-3c) is generally in the same direction as the motion of the moving object. However, both of these force vectors are applied parallel to the contact surface.

If you don't quite fully understand friction at this point, don't worry! I explain it in a whole lot more detail in Chapter 24.

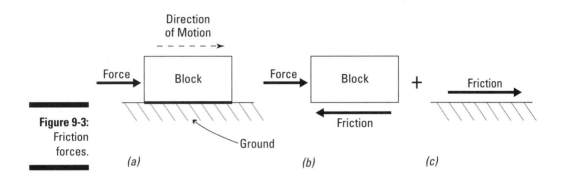

Figure 9-3:
Friction
forces.

(a) *(b)* *(c)*

Concentrated loads

A few concentrated loads that you want to remember are moments and support reactions. Both of these concepts play significant roles in your application of statics.

- ✔ **Moments:** *Moments* are physical effects that cause an object to want to rotate. In Chapter 12, I show you how to draw and calculate a moment.

- ✔ **Support reactions:** *Support reactions* are external conditions that restrain an object from moving and/or rotating in a specified direction. In Chapter 13, I show you how to draw and represent support reactions.

Other unique behaviors can result in concentrated loads, but these cases are special. I address those as I come across them throughout the book.

Revealing the Unseen with Concentrated Internal Loads

Internal loads are created through the application of external forces and effects and are always present on rigid bodies. You only see them when an object is physically cut. You can classify internal forces into three major categories: internal axial loads, internal shear forces, and internal moments.

- ✔ **Internal axial loads:** *Internal axial loads* are the simplest of the internal forces and are present in many objects, such as ropes and cables, simple columns, and springs. Internal axial loads are concentrated forces that act parallel to the longitudinal axis of the object. The senses of internal axial loads are always referred to as either *tension* (getting longer) or *compression* (getting shorter).

- ✔ **Internal shear loads:** *Internal shear loads* are concentrated forces that act perpendicular to the longitudinal axis of the object.

✔ **Internal moments:** *Internal moments* are actions that cause the object to rotate around a point or an axis and always appear when internal shear loads are developed.

Don't worry too much about shear and moments at this point; I discuss those in much more detail in Chapter 20. In this section, I focus on internal axial loads.

Figure 9-4a shows a bar subjected to tensile forces (P) applied on each end. This bar is perfectly balanced by these two forces acting in equal but opposite directions. If you cut the bar into two pieces and look at the lower portion (shown Figure 9-4b), you see that a new force, (P_{INT}) is developed inside the bar in order to help hold that portion of the bar in its original position. In order to hold the bar in place, P_{INT} must equal the applied load P. Likewise, if you look at the upper portion of the bar (shown in Figure 9-4c) by itself, another internal force is created to keep this part balanced as well. This internal force P_{INT} is also equal to the applied load P.

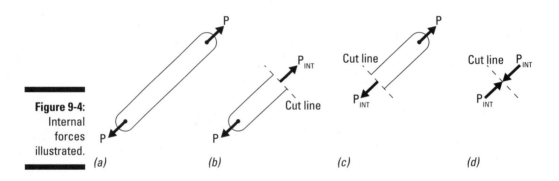

Figure 9-4:
Internal
forces
illustrated.

(a)   *(b)*   *(c)*   *(d)*

If you examine the direction of the same force P_{INT} on the upper and lower portions, you see that each of the internal forces are of the same magnitude P but have different directions depending on which side of the cut line you're working with (see Figure 9-4d). In Chapter 13, I show you how to better represent these internal forces, and I show you how to start writing equations and working with them beginning in Part VI.

Forces in ropes and cables

Forces in ropes and cables are always in an axial direction, along the line of the rope or cable, and are always in tension. These forces act along the line of action of the rope or cable. That is, the position vector of the rope is in the same direction as the unit vector of the force in the cable. For these types of objects, you often make use of dimensional data and coordinates.

In Figure 9-5, a force of magnitude 150 pounds is applied to the end of the rope. Because you know the location on the wall is a height of 30 feet, the Cartesian coordinates of the knot on the wall are (0,30,0). (Flip to Chapter 5 for more on Cartesian coordinates.) The end of the rope is located at coordinates of (10,10,10).

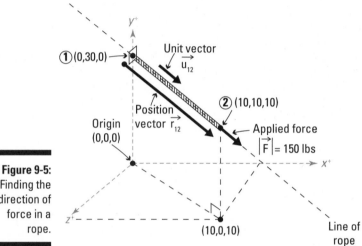

Figure 9-5:
Finding the direction of force in a rope.

In Chapter 5, I also mention that you can use a unit vector to define the direction of a line of action. In the case of these rope problems, you always know the line of action because the orientation of the rope itself is on that same line of action.

To write a force vector for the force on the rope, follow these steps:

1. **Create a position vector between the ends of the rope.**

   Define the end of the rope at the wall as Point 1 and the free end of the rope as Point 2. You can calculate the position vector that defines the rope from Point 1 to Point 2 from the basic position vector equation:

   $$\vec{r_{12}} = (x_2 - x_1)\hat{i} + (y_2 - y_1)\hat{j} + (z_2 - z_1)\hat{k}$$

   Point 2 is the end point of the position vector (the head) and Point 1 is the start point of the position vector (the tail); substitute the coordinate values from this example to produce a position vector for the ends of the rope:

   $$\vec{r_{12}} = (10-0)\hat{i} + (10-30)\hat{j} + (10-0)\hat{k} = 10\hat{i} - 20\hat{j} + 10\hat{k} \text{ ft}$$

2. **Calculate the magnitude of the position vector from Step 1.**

   To calculate the magnitude of a three-dimensional vector, you utilize the Pythagorean theorem, as I discuss in Chapter 5.

   $$\left|\overrightarrow{r_{12}}\right| = \sqrt{(10)^2 + (-20)^2 + (10)^2} = 24.495 \text{ ft}$$

3. **Use the position vector from Step 1 and the magnitude from Step 2 to calculate the unit vector that describes the line of action of the rope.**

   After you have the position vector and the magnitude (or length of the rope), you can create a unit vector by dividing the vector itself by its magnitude (see Chapter 5).

   $$\overrightarrow{u_{12}} = \frac{10\hat{i} - 20\hat{j} + 10\hat{k} \text{ ft}}{24.495 \text{ ft}} = 0.408\hat{i} - 0.816\hat{j} + 0.408\hat{k}$$

   The following quick check verifies that this result is indeed a unit vector:

   $$\left|\vec{u}_{12}\right| = \sqrt{(0.408)^2 + (-0.816)^2 + (0.408)^2} = 0.999\ldots \approx 1.0$$

   $\mathbf{u}_{12}$ as created from the dimensional data of ropes and cables always defines the same line of action as the line of action of the force, or

   $$\overrightarrow{u_{12}} = \overrightarrow{u_F} = 0.408\hat{i} - 0.816\hat{j} + 0.408\hat{k}$$

4. **Create the force vector by plugging the information gathered into the force vector formula.**

   Here's what your example looks like in the formula:

   $$\overrightarrow{F} = \left|\overrightarrow{F}\right| \cdot \overrightarrow{u_F} = (150 \text{ lbs}) \cdot \left(0.408\hat{i} - 0.816\hat{j} + 0.408\hat{k}\right)$$

   or

   $$\overrightarrow{F} = 61.2\hat{i} - 122.4\hat{j} + 61.2\hat{k} \text{ lbs}$$

   which defines the force acting on the end of the rope at (10,10,10) as a Cartesian vector.

   This vector notation also includes additional useful information. For example, based on your final calculation, you now know that the rope is experiencing a force of +61.2 pounds parallel to the positive x-direction, –122.4 pounds parallel to the positive y-direction (or +122.4 pounds in the negative y-direction), and +61.2 pounds parallel to the positive z-direction. These are the *Cartesian components* of the force vector **F** (which I discuss in more detail in Chapter 8).

5. Verify that your force vector is correct by verifying its magnitude.

To verify the magnitude, you just use the Pythagorean theorem a second time and plug in the component values for the x-, y-, and z-components.

$$|\overline{\mathbf{F}}| = \sqrt{(61.2)^2 + (-122.4)^2 + (61.2)^2} = 149.9 \text{ lbs} \approx 150 \text{ lbs}$$

Notice that the exact value of the magnitude is only approximately equal to 150 pounds because the scalar force components were only taken to one decimal place.

Forces in springs

Another type of axial force that you encounter is a *mechanical spring object*. When you think of a spring, you probably picture that wonderful toy you owned as a child and the countless hours you spent flopping it down your stairs (or trying to straighten it out). In statics, springs are a bit different. They may resemble toy springs, but they actually behave quite differently. Figure 9-6a illustrates a common depiction of a spring object.

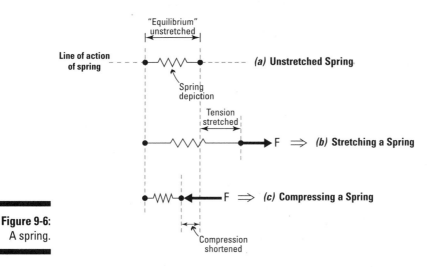

Figure 9-6:
A spring.

Several features are very important when discussing springs, including line of action, stretch, and spring constants:

✔ **Line of action:** As with the depiction of force vectors (see Chapter 4), the *line of action* of a spring refers to the direction of the longitudinal axis. The two points that connect the spring to other objects or supports always occur along the line of action of the spring. As a result, the internal axial force of the spring is always along this line.

✔ **Stretch:** The *stretch* refers to how much the spring is displaced from its original unstretched length. Stretching a spring to make it longer is tension or *elongation* (shown in Figure 9-6b), whereas stretching it to make it shorter is called *compression* (Figure 9-6c). The more you stretch a spring, the more force is developed in the spring.

✔ **Spring constants:** The *spring constant* is a measure of how much force you need to compress or elongate a spring from its unstretched position. Factors that affect spring constants include the cross-sectional area of the spring, the unstretched length of the spring, and the material the spring is made of.

The following sections look at stretch and spring constants; for more on line of action, check out Chapter 4.

Stretch in springs

The force in a spring is directly related to the amount and direction of the stretch (or *deformation*) of the spring. A negative stretch indicates a compressive force; a positive stretch denotes an elongating force. You can compute the stretch (or *spring deformation*) from

$$\delta_x = \text{final length} - \text{unstretched length}$$

Spring constants

Spring constants define how stiff a spring actually is and are a measure of how much force is necessary to stretch (or compress) a spring a distance of one unit. In SI units, the unit of the spring constant is usually Newton per meter (N/m), and in U.S. customary units, it's pounds per inch (lb/in) or pounds per foot (lb/ft). For example, Figure 9-7 depicts a spring with an unstretched length of 6.5 inches and a spring constant of 3,000 pounds per inch. A force applied to the spring compresses the spring to a final length of 5.25 inches.

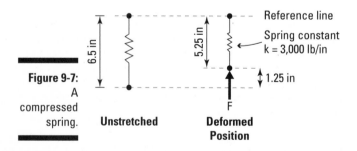

Figure 9-7:
A
compressed
spring.

To compute the stretch δ_x,

$$\delta_x = \text{final} - \text{unstretched} = 5.25 - 6.5 = -1.25 \text{ inches}$$

The equation that represents the magnitude of the internal force in a spring is given by

$$\left|\vec{F}_{SPRING}\right| = k \cdot \delta_x$$

where k represents the spring constant parameter and δ_x represents the spring's deformation from its unstretched state. You can compute the force in the example spring by plugging in the numbers:

$$\left|\vec{F}_{SPRING}\right| = \left(3{,}000\,\frac{\text{lb}}{\text{in}}\right) \cdot \left(-1.25 \text{ in}\right) = -3{,}750 \text{ lbs} = -3.75 \text{ k}$$

The negative sign on the force indicates that the force on Figure 9-7's spring is a compressive force. This finding supports the idea that a shorter spring is a compressed spring.

If you want to determine the direction of the force, you can create a unit vector for the line of action as I demonstrate in the "Forces in ropes and cables" section earlier in this chapter.

Surveying Self Weight as an External Load Value

For many people, self weight is an ominous concept, but don't worry — you're not climbing on any scales here. In statics, the *self weight* of an object is a measure of the force created by gravity's effects on the object's mass.

- ✔ **Gravity:** *Gravity* is a force of attraction between masses or particles. On Earth, at sea level, the average gravitational acceleration constant is taken as 9.81 meters per second squared (m/s^2, SI units) or 32.2 feet per second squared (ft/s^2, U.S. customary units). These values actually vary slightly by location because the Earth isn't a perfectly round sphere. However, this variation is very minor, so I use these average values throughout the book.

- ✔ **Mass:** *Mass* is a measure of the number of atoms in an object in conjunction with the density of each of those atoms; the mass of an object is a fundamental property of that object. Mass is measured in kilograms (kg) in SI units and in slugs ($\text{lb-s}^2/\text{ft}$) in U.S. customary units.

Self weight is usually either treated as a single concentrated value, which I discuss in this section, or spread over a continuous region, which I discuss in Chapter 10. Regardless, the basic formula needed to calculate the self weight, W of an object is given by

$$\overrightarrow{W} = m \cdot \overrightarrow{g}$$

where *m* represents the mass of the object and g represents the acceleration due to gravity. In this form, the weight W of an object is a vector force in the direction of the gravitational acceleration g. On Earth, you always assume that gravity is acting locally downwards. (Of course, if you're standing on your head, your sense of up and down may be a bit skewed.) Gravity always acts towards the center of the Earth. I discuss more about the location of application of this force in Chapter 11.

Getting specific on specific gravity and self weight properties

When working with self weight, you may sometimes encounter problems that don't directly state the mass or the weight of an object. However, you may be able to use other terms, such as the following, to calculate those figures:

- ✔ **Density:** The *density* of an object is a measure of the amount of mass of an object contained within a certain volume of that object. For example, dropping a bowling ball on your foot is a lot different than dropping an air-filled balloon of the same size on your foot, and that's because of the effects of density. Temperature is a major factor in the density of an object. The density of an object is measured in units of kilograms per cubic meter (kg/m^3) for SI units, in $lb\text{-}s^2/(ft\text{-}ft^3)$ (I won't bore you with the spelled-out version of that one) in U.S. customary units. Density is usually expressed by the Greek variable ρ, or rho.

- ✔ **Specific gravity:** *Specific gravity* (sg) is the density of the material relative to the density of water, which has a specific gravity of 1.0. One useful feature of specific gravity is that the value is unitless and remains the same regardless of which measurement system you use. So a specific gravity of 1.0 in the metric system is a specific gravity of 1.0 in the U.S. customary system.

  For example, carbon steel has a specific gravity of approximately 7.8, meaning that it's nearly eight times as heavy as the same volume of water. Ice, on the other hand, has a specific gravity of approximately 0.92 (which means that it's lighter than water, and one of the reasons that it floats!).

If the problem gives you the weight of the object, you must include it in your calculations. If the mass, density, or specific gravity is provided, you can calculate the weight from the formulas presented in this section. In most structural problems, the applied forces on an object are often much, much greater in magnitude than the actual self weight of the object. So, if you choose to neglect the self weight when the applied loads are large, you'll probably be okay. However, keep in mind that if you ignore the self weight of an object weighing 200 pounds when the applied loads are only 5 pounds, your results will end up being highly inaccurate.

Specific weight (γ) is the weight per unit volume and is a relationship that relates density (mass per unit volume) of a material with gravitational effects. The formula is:

$$\gamma = (\rho) \cdot (g)$$

where γ is the specific weight, measured in Newton per cubic meter or pounds per cubic foot in SI and U.S. customary units, respectively.

Working with lumped mass calculations

Mass is generally distributed throughout each particle of an object. However, instead of calculating each weight for each particle (and you may have a *lot* of particles) in an object, you can simplify your work with mass and weight by making an assumption about the location at which the weight or mass is acting. Compute the grand total of all the particle masses and then express this value as a single value, or *lumped mass*. You can compute the lumped mass of an object, *m,* from the following relationship:

$$m = \rho \cdot V$$

where ρ is the density of the object (as defined earlier in this chapter), and *V* represents the volume of the object measured in cubic meters or cubic feet in SI and U.S. customary units, respectively. So, if you know the density of a cube of water is approximately 1,000 kilograms per cubic meter at 60 degrees Fahrenheit and that the cube has a volume of 0.35 cubic meters, you can calculate the mass of that water cube as

$$m = \left(1{,}000 \frac{\text{kg}}{\text{m}^3} \right) \cdot \left(0.35 \text{ m}^3 \right) = 350 \text{ kg}$$

This method is valid for *prismatic* objects, or simple objects having constant dimensions in each direction.

Introducing the Principle of Transmissibility

The *principle of transmissibility* implies that a force vector acting on a rigid body results in the same behavior regardless of the point of application of the force vector, as long as the force vector is applied along the same line of action. It's a concept that's very important within vector mechanics. In fact, it forms the basis for one of the three major assumptions that Isaac Newton proposed with respect to rigid body mechanics — it doesn't affect the calculations for equations of equilibrium (covered in Chapter 16) because ultimately, the direction, point of application, and magnitude of the applied vector are still the same. Figure 9-8 shows a graphic representation of the principle of transmissibility for rigid bodies.

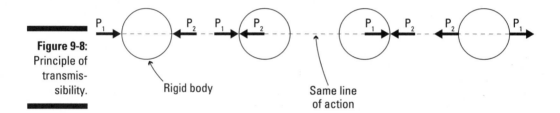

Figure 9-8: Principle of transmissibility.

Rigid body

Same line of action

Notice that for the rigid body shown, a force P_1 acting to the right and another force P_2 acting to the left result in the same net behavior as long as the forces maintain their original magnitude and sense, and act on the object along the same line of action. That is, the point of application of forces can occur anywhere along the same line of action on the object without changing the resulting behavior of the object.

Chapter 10

Spreading It Out: Understanding Distributed Loads

In This Chapter
▶ Defining the properties and types of distributed loads
▶ Finding a distributed load's resultant
▶ Distributing mass throughout an object
▶ Seeing the connection between concentrated and distributed forces

When you throw a small object such as a golf ball at a wall, the force the ball makes onto the wall (and of course the force that the wall makes onto the ball) acts on a very small area. In many cases, you idealize this force as a concentrated load (which I cover in Chapter 9) because the force is concentrated onto a small area. But what happens when you throw a larger object (such as your malfunctioning TV) with the same force at the same wall? In this case, the area of the force when it meets the wall is spread out and is therefore no longer concentrated. This type of force is known as a distributed force and has several unique properties that you need to remember.

In this chapter, I discuss distributed loads in detail and show you how to calculate their total combined effect. I also reveal how you can consider self weight (see Chapter 9) as a distributed load as well.

Getting a Handle on Some Distributed Load Vocab

A *distributed load* is a load that doesn't act at a point but rather is spread out over a specified length. Because distributed forces are spread out over the entire section, they have some properties you want to be aware of: intensity, start point, and end point.

✔ **Intensity:** The *intensity* of a distributed load is actually the *magnitude* of the load (flip to Chapter 4 for more on magnitude). For a distributed load, intensity can be any shape. It's often a polynomial of zero, first, or second order (see Chapter 2); however, you can use any function to describe the intensity depending on the loading you're describing. In the following section, I describe several common shapes of intensity functions that you may encounter.

✔ **Start point:** The *start point* of a distributed load indicates where the intensity load application actually begins.

✔ **End point:** The *end point* of a distributed load indicates where the intensity loading actually ends.

Figure 10-1 shows a depiction of a fully distributed load (where the start point is at one end of the object and the end point is at the opposite end), a partially distributed load, and a concentrated load applied to a beam.

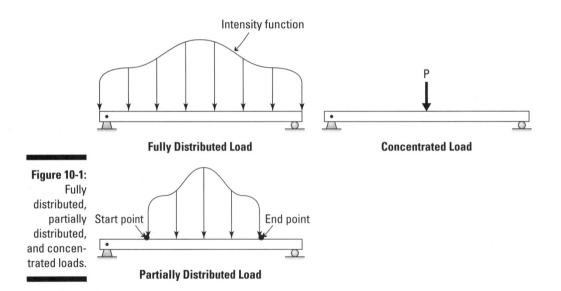

Figure 10-1:
Fully distributed, partially distributed, and concentrated loads.

Fully Distributed Load

Concentrated Load

Partially Distributed Load

Distributed loads come in many shapes and varieties depending on how the load was created and how it's applied on the object. Distributed loads can be linear, surface, or volumetrically distributed; I deal with these varieties in more detail in the following section.

Take a (Distributed) Load Off: Types of Distributed Loads

You can typically classify distributed loads based on the dimensions in which they're acting. The units of a load intensity always give you some insight into the type of load you're dealing with and can help you choose an appropriate method for working with distributed loads. Check out the following sections for more on the types of loads and the units they use.

Distributed forces

For *distributed forces* that act over a linear distance, you measure in force-per-distance measurements: Newton per meter (N/m) for SI units and pounds per foot (lb/ft or plf) in U.S. customary units. Figure 10-2 shows several different distributed forces. By far, the two most common of the distributed forces are known as uniformly distributed and linearly distributed (though they're not the only options).

- ✔ **Uniformly distributed:** A *uniformly distributed load* is a load that has a constant intensity over the length of the load. Figure 10-2a illustrates a uniformly distributed load with a constant magnitude of w_o. A uniform load is also a zero-order load distribution because the order of a polynomial of zero order ($n = 0$) for a polynomial is of the form $w(x) = w_o x^0 = w_o$, which indicates a zero-order (or constant) function.

- ✔ **Linearly distributed:** A *linearly distributed load* is a load with an intensity that varies linearly over the length of the load. The lowest intensity occurs at one end of the load, and the maximum occurs at the other. You can establish all intensity values in between from a linear function, as shown in Figure 10-2b. In a linearly varying function, the function is first order because it has the form $w(x) = ax+b$, which defines a linear function.

- ✔ **Higher order distributed:** A *higher order distributed load* is a distributed load that has an intensity that can be determined by a polynomial of order greater than one (or $n > 1$) or by completely different functions altogether (such as trigonometric functions). Higher-order distributions such as the one shown in Figure 10-2c can be quite complex and may require special calculations.

Surface loads (pressures)

Surface distributed loads (also known as *surface pressures*) are loads that act over a prescribed area. Over that area, the intensity of a surface load can vary greatly. You measure a surface distributed load as a force per area, so you express the units as Newton per square meter (N/m^2) or the pascal (Pa.) in SI and pounds per square foot (lb/ft^2 or psf) in U.S. customary.

Forces spread over an area are also called pressures. Pressures always act over a two-dimensional area (or surface) and can vary in intensity in multiple directions. Examples of surface pressures include the pressure of water acting on a dam (pressures are very common when dealing with fluids), and the weight of snow on the roof of your house. Figure 10-3 shows a pressure on a surface; the pressure intensity has the form $w(x,y)$ because it can have changing intensities in the x- and y-directions.

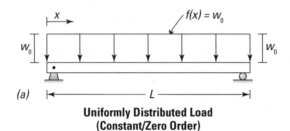

(a)

Uniformly Distributed Load
(Constant/Zero Order)

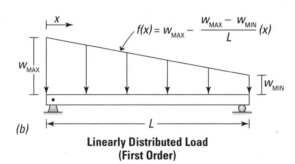

(b)

Linearly Distributed Load
(First Order)

Figure 10-2:
Distributed
forces.

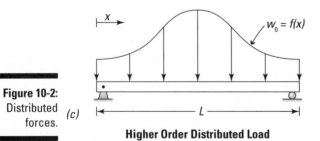

(c)

Higher Order Distributed Load

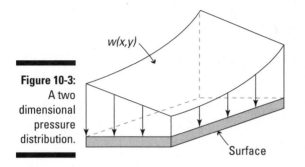

Figure 10-3:
A two
dimensional
pressure
distribution.

w(x,y)

Surface

Pressure loads are also used frequently in structural analysis to model room capacities of people. Various design codes require anywhere from 40 to 100 pounds per square foot of design load depending on the size, occupancy, and purpose of a particular room.

Volumetric loads

Volumetric distributed loads are loads that act over a volume. The most common volumetric load is the *specific weight,* a useful value for calculating the self weight of an object. (Take a look at Chapter 9 for details on specific weight and self weight.) You measure a volumetric distributed load as a force per volume, so you use the units Newton per cubic meter (N/m^3) for SI and pounds per cubic foot (lb/ft^3) in the U.S. customary system. Volumetric loads require all three dimensions to calculate (versus the two dimensions pressures require; see the preceding section).

Calculating the Resultant of a Distributed Load

Sometimes in statics work, you want to be able to consolidate a distributed load into a single value that acts at a single, specific location. This combined or consolidated force is known as the *resultant.* The resultant of a distributed load is similar to the resultants of concentrated loads I discuss in Chapter 7. With distributed loads, the resultant is a single combined force that represents the entire effect of the distribution.

After calculating the resultant, you still need to find the unique point of application where it's acting. I show you how to compute this location in Chapter 11.

Uniform and linearly varying forces

The simplest resultant forces to calculate are those forces that occur in two dimensions. In two dimensions, you calculate the resultant of a distribution by evaluating the following relationship: resultant = area under loading distribution.

To start the calculating process, I like to replace the load diagram with a dashed line to represent the boundary as shown in Figure 10-4. In addition, I place a concentrated force to represent the resultant of the distributed load. The dimension x helps determine the specific location of the resultant, which I show you how to calculate in Chapter 11. I use this type of sketch for two reasons. First, it helps determine the necessary dimensions of the area calculation. Second, it also preserves a reminder of the original load distribution that I use to compute the resultant magnitude.

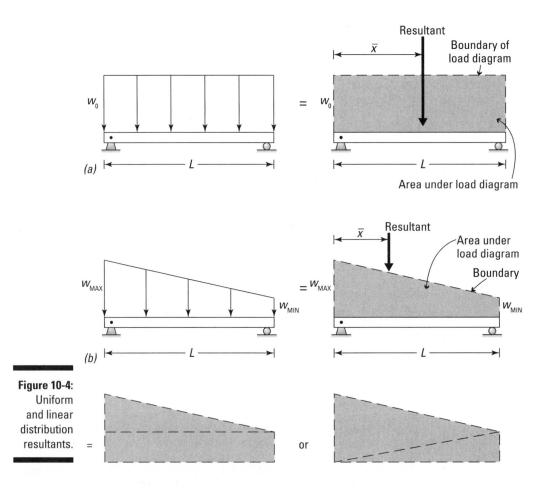

Figure 10-4:
Uniform and linear distribution resultants.

Zero order (uniform) distributions

The uniform distribution in Figure 10-4a is one of the simplest resultants to compute. By looking at the distribution boundary line, you see that the resultant area that you need to compute is actually the area of the rectangular shaded region.

Recall from your basic geometry class (remember that?) that the formula for the area of a rectangle is area = base · height, where base and height are the dimensions of the rectangle. You can expand this formula to apply to uniform distributions:

$$\text{Resultant} = \text{Area}_{\text{RECTANGLE}} = \text{base} \cdot \text{height} = w_{\text{o}} \cdot L$$

For the uniform distribution, the base is the length of the length of the uniform distribution, or L. The height of the distribution is actually the intensity, w_{o}, of the distribution.

For example, if you stack 300 pounds over a one-foot area and then copy this loading repeatedly for the entire length of the beam, you've just defined a uniform load of 300 pounds per foot. If the beam is 12 feet long, the resultant load of this uniform distribution is 300 pounds per foot · 12 feet = 3,600 pounds.

Thus, the resultant of this uniform load is 3,600 pounds total acting on the beam.

First order (linearly varying) distributions

The linearly varying distribution shown in Figure 10-4b is another common distribution that you often encounter.

To compute the magnitude of the resultant, you need to be able to compute the area of a trapezoid (the shaded area of the distribution's boundary):

$$\text{Resultant} = \text{Area}_{\text{TRAPEZOID}} = \frac{1}{2}\left(w_{\text{MAX}} + w_{\text{MIN}}\right) \cdot \left(L\right) = \frac{1}{2}\left(w_{\text{MAX}}\right) \cdot \left(L\right) + \frac{1}{2}\left(w_{\text{MIN}}\right) \cdot \left(L\right)$$

where length of the distribution is given by the dimension L, the maximum intensity is given by w_{MAX}, and the minimum intensity is given by w_{MIN}. Geometrically, you can represent this calculation as a sum of a rectangle and triangle or as a sum of two triangles as shown in Figure 10-5 in the following section.

Other two dimensional distributions

Other two-dimensional distributions that you may encounter can be described by higher-order continuous or trigonometric functions and distributions composed of combinations of simple distributions.

Figure 10-5 illustrates a more complex two-dimensional distribution that has a varying intensity, $w(x)$, acting over a specified length L.

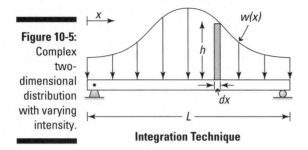

Figure 10-5:
Complex two-dimensional distribution with varying intensity.

Integration Technique

If you remember from calculus, you can easily use an integral to determine the area bound by a continuous function. You start by taking an incremental slice that has a differential width, at some arbitrary position x. The height of the slice is actually the value of the intensity function, $w(x)$, evaluated at that point. That is, $h = w(x)$. Using integration, you can then compute the area from the following integral:

$$\text{Resultant} = \text{Area} = \int_A dA = \int_0^L h \cdot dx = \int_0^L w(x) \cdot dx$$

REMEMBER

As long as you can define the intensity of the distribution as a function $w(x)$, or a series of functions $w_1(x)$, $w_2(x)$, and so on, you can compute the magnitude of the resultant of the distribution.

For working with complex or combinations of distributions, you can always break up a larger distribution into smaller (and often simpler) pieces. Figure 10-6 illustrates a distribution that's actually a combination of two simpler distributions.

In this example, you can break this seemingly complex distribution into a combination of a linearly varying distribution shown as Area #1 and a uniform distribution shown as Area #2, and you can compute the resultant for each of those two distributions separately. After you determine these resultants, you can then calculate the resultant of the two smaller resultants (Resultant #1 and Resultant #2). This final resultant then has its own position, as indicated by x_{RES}. I discuss this topic more in Chapter 11.

Surface loads and pressures in multiple dimensions

In concept, calculating the resultant of a surface load is similar to how you treat linear distributed loads. Instead of calculating the area under a linear distribution, for surface loads, you actually calculate the volume inside the pressure distribution. Figure 10-7 shows an example of a three-dimensional pressure diagram.

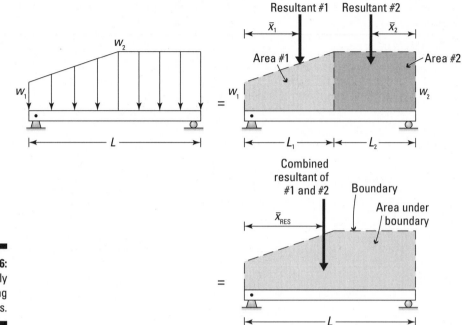

Figure 10-6:
Two linearly
varying
distributions.

For a given Cartesian axis, you know (or can determine from calculus) the function that describes the pressure distribution. In Figure 10-7, the pressure varies in two directions (both x and y) — that is, a pressure distribution can vary in two directions. To find the resultant, you have to use some basic calculus involving a double integral for this evaluation:

$$\text{Resultant} = \text{Volume}_{\text{DIST}} = \int_0^h \int_0^L p(x,y) \cdot dx \cdot dy$$

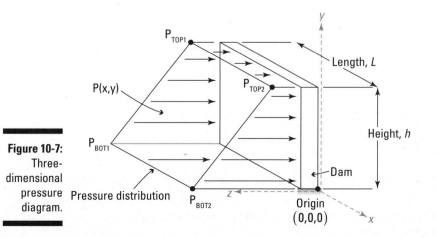

Figure 10-7:
Three-
dimensional
pressure
diagram.

Now, before you completely panic about all the calculus you may have forgotten, here's how you can handle a pressure distribution defined by a geometric prism. A *geometric prism* is a three-dimensional shape that has equal areas on opposite ends of the shape and a constant length. In Figure 10-7 earlier in the section, if $P_{TOP1} = P_{TOP2}$ and $P_{BOT1} = P_{BOT2}$, the distribution shown is a geometric prism. To calculate the volume of this prism, you can make use of a basic equation for calculating the volume of a prism:

$$\text{Volume} = \text{Area}_{BASE} \cdot \text{height}$$

Avoiding the double integral

If you redraw the distribution of Figure 10-7 as a two-dimensional linear distribution that is constant along its length, you can create a new, simpler distribution as shown in Figure 10-8. If you use the following basic steps, you can avoid the double integral and calculus.

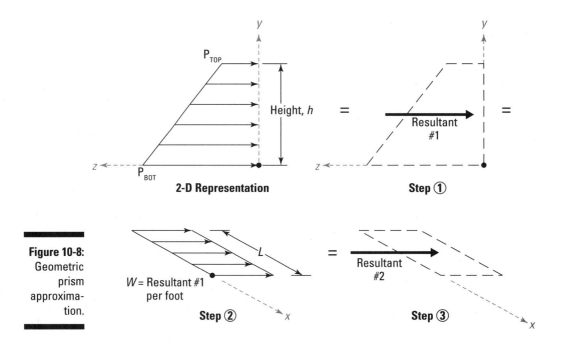

Figure 10-8: Geometric prism approximation.

1. **Determine the resultant (Resultant #1) of the distribution for the cross-sectional area of the geometric prism, using the techniques I discuss in the section "Uniform and linearly varying forces in two dimensions."**

   When you calculate this resultant, you end up with a force per distance, so your units are Newton per meter or pounds per foot.

2. **Create a new distribution in the direction of the length of the original distribution.**

   As shown in Figure 10-8, you evaluate the second distribution in the Cartesian x-direction. The intensity of this new distribution is equal to Resultant #1 and is applied uniformly along length L.

3. **Use those same techniques to resolve this second distribution into a final resultant (Resultant #2).**

   Resultant #2 is actually the resultant of the surface pressure. The units on this final resultant are either Newton or pounds.

This technique only works on distributions that can be defined as geometric prisms. If you can't define the shape as a geometric prism, you have no choice but to use the double integrals technique.

Looking at Mass and Self Weight as Distributed Values

As I state in Chapter 9, treating mass and weight as a single lumped value is valid for *prismatic* objects, or simple objects having constant dimensions in each direction. However, for cases where the object shape is irregular (non-prismatic), or the density of the object isn't constant throughout, the mass may have a distribution of its own. For example, if you take a prismatic bar that has a constant density and cut it into two pieces, each piece still has a weight associated with it, which means each piece must have mass as you can see in Figure 10-9.

Figure 10-9 shows a prismatic bar with a length L that has been broken into n equal-length pieces. That means that the length of each piece is given by $\Delta L = \frac{L}{n}$.

If the cross section of the bar is not prismatic, you have to assume that the mass is constant for any given piece and calculate the mass (as I describe in Chapter 9) for each piece separately. A distributed mass problem is basically a lot of lumped mass problems combined into a single problem.

After you determine the mass of each piece, you can calculate the self-weight $(W)_i$ of each piece by using the following formula:

$$(W)_i = \left(\frac{\text{mass}}{\Delta L} \right)_i \cdot g$$

where g is the acceleration due to gravity.

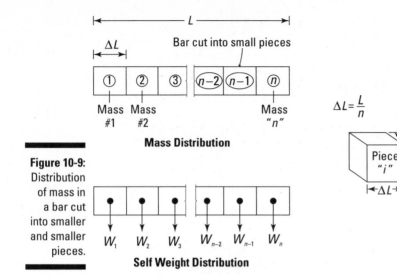

Figure 10-9:
Distribution
of mass in
a bar cut
into smaller
and smaller
pieces.

Each piece of the bar actually has its own self weight. Usually you can assume that the mass is constant over very small increments (or on a piece-by-piece basis), but from one piece to the next, the distributed masses can be vastly different.

Here's a handy guide to the units and constants you need (in both SI and U.S. customary) to work with distributed mass:

Measurement	SI	U.S. Customary
Mass	kg/m	slugs/ft
Gravity	9.81 m/sec^2	32.2 ft /sec^2
Weight	N/m	plf

To verify the piecewise values, you can still determine the total weight of the bar by adding the weights of all the individual pieces:

$$W_{\text{TOTAL}} = \sum_{i=1}^{n} \left[(W)_i \cdot \Delta L_i \right]$$

The final units of W_{TOTAL} give you the same units (Newton for SI and pounds for U.S. customary) as the lumped mass system in Chapter 9.

Chapter 11

Finding the Centers of Objects and Regions

*I*n Chapter 10, I describe how to find the *resultant* (combined) force of a variety of distributed loads by simply determining the areas under the load diagram. This calculation provides you with two of the three pieces of information required to fully define a force vector — namely, the *magnitude* (the vector's length) and the *sense* (the vector's direction). However, you also need to determine a force vector's point of application in order to properly define the vector. (Check out Chapter 4 for details on these vector properties.)

For *concentrated loads* (single loads applied at a point — see Chapter 9), you can determine the point of application almost by inspection. If a small object hits a wall, a concentrated force from the ball is located at the point of impact. However, *distributed loads* (loads spread over a line or area — see Chapter 10) are different.

To find the point of application of a resultant of a distributed load, you have to calculate the center of area or the centers of mass and gravity for the load or object. In this chapter, I show you how to perform these calculations.

Defining Location for Distributed Loads

Depending on the type of distributed loads you encounter along your statics journeys, the resultant force of each of those loads must act at a specific location. As you understand how to specify the locations where distributed loads and *self weight* (the force resulting from the gravitational effects of the mass of an object) are positioned, you encounter several terms to define these positions: centroid, center of mass, and center of gravity.

- ✔ **Centroid:** The *centroid* or *center of area* of a geometric region is the geometric center of an object's shape. For most *external* distributed loads (where the force acts on the object from the outside), the resultant force acts at a location known as the centroid of the load distribution. See the following section for more on centroids.

- ✔ **Center of mass:** The *center of mass* is the location at which the resultant mass is assumed to act.

- ✔ **Center of gravity:** The *center of gravity* is the average location at which the self weight of the object is assumed to act. Usually, the center of gravity and center of mass are assumed to be the same location; I explain why a little later in this chapter.

Chapter 9 gives you the lowdown on lumped self weight, and Chapter 10 describes distributed self weight.

Getting to the Center of Centroids

The centroid is actually a set of coordinate values (x,y,z) measured relative to a specific reference point. Usually, the origin — or coordinates (0,0,0) — of the Cartesian coordinate system that you implement is a convenient reference point. (See Chapter 5 for more on Cartesian coordinates.) For many shapes, this location often occurs inside the boundaries of the region. However, in some situations you may actually compute the centroid coordinates at a position outside the boundary.

Determining the location for the resultant of a distributed load involves calculating the centroid of the load region, which I show you how to do in the following sections.

Defining a centroid's region type

When you compute a centroid location, your first step is to always determine which equation you should use. To make this decision, you must first classify whether a region is discrete or continuous.

- ✔ **Discrete region:** A *discrete region* is an area that can be broken up into several subregions composed of simple shapes, such as rectangles, circles, triangles, and parabolic segments, with known or easily determined areas. You can also easily express the centroids of these regions based on the dimensions of the region.

✔ **Continuous regions:** A *continuous region* is any region that isn't classified as a discrete region. This region is normally enclosed by a complex or irregular-shaped boundary. To determine the centroid of continuous regions, you have to define the boundaries by using mathematical functions and then employ basic calculus and integration techniques.

Computing the centroid of a discrete region

A *discrete region* is a type of region made up of a combination of shapes referred to as *subregions*. Each subregion has its own individual area and centroid calculation that is usually fairly simple to compute. You can then combine these subregion properties to compute a single centroid location by using the following equations:

$$\bar{x} = \frac{\sum_{i=1}^{n} x_i A_i}{\sum_{i=1}^{n} A_i} \quad \text{and} \quad \bar{y} = \frac{\sum_{i=1}^{n} y_i A_i}{\sum_{i=1}^{n} A_i}$$

✔ $\bar{x}$ is the distance from the origin to the centroid of the discrete region measured parallel to the Cartesian x-axis.

✔ x_i is the distance from the origin to the centroid of subregion *i* measured parallel to the Cartesian x-axis.

✔ $\bar{y}$ is the distance from the origin to the centroid of the discrete region measured parallel to the Cartesian y-axis.

✔ y_i is the distance from the origin to the centroid of subregion *i* measured parallel to the Cartesian y-axis.

✔ A_i is the area of subregion *i*.

✔ *n* is the number of subregions that make up the discrete region.

If the *sigma notation* in this equation looks foreign to you, flip to Chapter 2.

Noting geometric properties of simple shapes

Figure 11-1 shows six simple shapes that allow you to handle the centroidal calculations for the vast majority of discrete regions. With these six basic shapes, you can construct many more-complex discrete regions.

Pay special attention to the location of the origin from which the centroid distances x_i and y_i are determined. Most statics books and other reference sources include similar graphics for properties of areas (usually inside the front or back cover), but the authors of these texts may base their measurements on completely different origins.

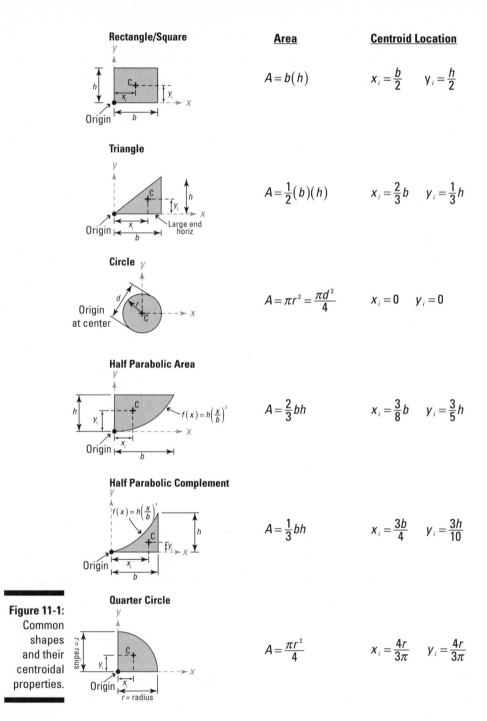

Figure 11-1: Common shapes and their centroidal properties.

Building a centroid calculation table

When calculating the centroid of discrete shapes, such as the one in Figure 11-2, I find that constructing a simple table makes the solution process much easier and more straightforward.

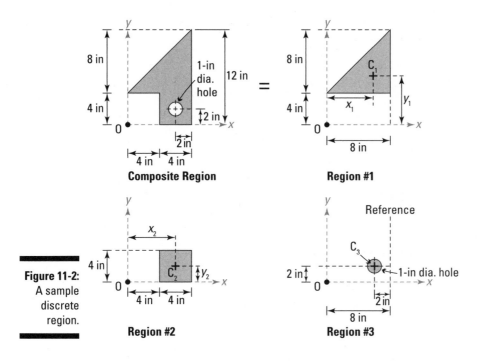

Figure 11-2:
A sample discrete region.

To calculate a centroid coordinate, you need a separate table for each x- and y-centroid dimension. To determine the x-centroid location, you start by creating a table with the column headings shown in Table 11-1.

Table 11-1		X-Centroid Coordinate Table	
Region #	*x_i (in)*	*A_i (in^2)*	*$x_i A_i$ (in^3)*
1 (triangle)	0.67(8) = 5.36	0.5(8)(8) = 32.00	(5.36)(32.00) = 171.52
2 (rectangle)	4 + 0.5(4) = 6.00	(4)(4) = 16.00	(6.00)(16.00) = 96.00
3 (circle hole)	8 – 2 = 6.00	–π(0.5)^2 = –0.785	(6.00)(–0.785) = –4.71
TOTAL	------------------	$\sum A_i = 47.22$	$\sum x_i A_i = 262.81$

Next, use the following simple steps to help you complete the table. At the end, the calculation in Step 7 is the actual x-coordinate location.

1. **In the first column, list each of the areas that make up the discrete region, including any holes or subtracted regions.**

2. **In the second column, calculate the distance from the origin of the combined region to the centroid of each shape.**

   For example, Region #1 ($i = 1$) is a triangle; Figure 11-1 earlier in the chapter shows you that the x-distance to the centroid of a triangle is

   $$x_1 = \frac{2}{3}b = 0.67 \cdot 8 = 5.36 \text{ inches}$$

3. **Calculate the area for each region and fill the results into the third column.**

   For Region #1, you can calculate the area of a triangle from

   $A_1 = \frac{1}{2}(8)(8) = 32.00$ square inches

   For regions that are missing or subtracted from other regions (such as holes), you calculate the area of the region containing the hole (Region #2) as though the hole doesn't exist and then calculate and subtract the area of the hole (Region #3). See "Including holes in discrete regions" later in the chapter for more information about this process.

4. **Multiply the values in the second and third columns and put the product in the fourth column.**

   For Region #1:

   $x_1 A_1 = (5.33)(32.00) = 171.52$ cubic inches

   Notice that this product for Region #3 contains a negative value, because the hole creates a negative area, as I note in Step 3.

5. **Add all the values in the third column and record this value on the bottom row for the TOTAL of that column.**

6. **Repeat Step 5 for the values in the fourth column.**

7. **Compute the $\bar{x}$ coordinate for the combined discrete region by dividing the total from Step 6 by the total from Step 5:**

   $$\bar{x} = \frac{262.81 \text{ in}^3}{47.22 \text{ in}^2} = 5.57 \text{ in}$$

Based on the result of Step 7, you can now locate the x-coordinate of the centroid, which is located 5.57 inches from the origin in the positive x-direction. You measure this distance from the same origin you use in all the calculations.

To locate the y-coordinate of the centroid, you need to create a table similar to Table 1-1, using y in place of x.

Including holes in discrete regions

When you're working with a strangely shaped discrete region, sometimes it's convenient to overestimate an area with a simple shape and then subtract another simple shape from your calculations. Using Figure 11-3a, you can overestimate the total region by drawing a rectangle with a horizontal dimension of $b_1 + b_2$ and a vertical dimension of $h_1 + h_2$. However, you're overestimating the total area of the actual region (see Figure 11-3b). To correct this estimation, you can then subtract a region having a horizontal dimension of b_2 and a vertical dimension of h_2 (see Figure 11-3c). The area of this subtracted region is computed as a negative number and included in Table 11-1 earlier in the chapter.

You can add or subtract regions from your estimated shape as long as you apply the correct sign to the area of the region when you calculate it. Areas being subtracted are always negative.

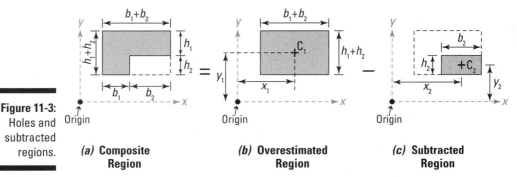

Figure 11-3: Holes and subtracted regions.

(a) **Composite Region**   *(b)* **Overestimated Region**   *(c)* **Subtracted Region**

You must measure the distance to the centroid of each simple area, including subtracted regions relative to the same reference point.

Handling trapezoidal regions

One of the more common discrete regions you come across is the *trapezoidal region,* which shows up frequently in submerged surface calculations (which I discuss in Chapter 23) and linearly varying load distributions (head to Chapter 10). Just like with other discrete regions, you can separate the trapezoidal region into smaller subregions. Figure 11-4 shows two possibilities for this division.

The first option is to break the trapezoid in Figure 11-4a into a rectangle and a triangle (see Figure 11-4b). The second option is to break the region into two triangles (see Figure 11-4c). Regardless of which method you choose, your centroid calculations produce the same resulting centroid location (assuming you do the math right, of course!).

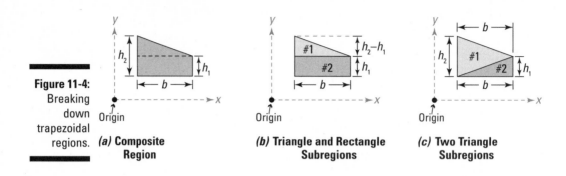

Figure 11-4:
Breaking
down
trapezoidal
regions.

(a) Composite
Region

(b) Triangle and Rectangle
Subregions

(c) Two Triangle
Subregions

Finding centroids of continuous regions

Finding the centroidal coordinates for a continuous region is usually more mathematically complex than the discrete centroidal calculations that I describe in the preceding section because you need to use your calculus skills to perform the integrations to find the centroidal coordinates. The equations that you need to use when working with continuous regions are

$$\bar{x} = \frac{\int_A x \cdot dA}{\int_A dA} \quad \text{and} \quad \bar{y} = \frac{\int_A y \cdot dA}{\int_A dA}$$

If you examine these formulas, the continuous region formulas are actually very similar to the discrete region formulas. To illustrate how these equations are used, consider the continuous region in Figure 11-5a, which is bounded by two functions, $f_1(x) = x^2$ on the lower bound edge and $f_2(x) = x$ on the upper bound edge.

To use these integral equations, you first need to develop expressions for the incremental area dA. Start by examining the shaded incremental slice shown. This rectangular area can be computed from

$$dA = h(x) \cdot dx = \left(f_2(x) - f_1(x)\right) \cdot dx = \left(x - x^2\right) \cdot dx \qquad \left(0 \le x \le 1\right)$$

The distance x in this centroidal calculation is the distance from the origin to the centroid of the rectangular incremental area. It's just the same variable x that you use when you perform your integration calculation.

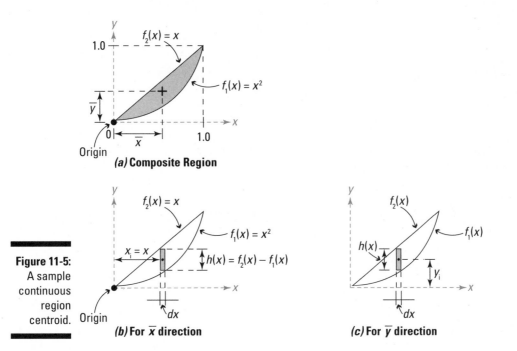

(a) Composite Region

Figure 11-5:
A sample
continuous
region
centroid.

(b) For $\overline{x}$ direction

(c) For $\overline{y}$ direction

Because you've now transformed the area integration into a linear integration calculation (as indicated by the dx), you need to change the limits of integrations as well. The upper limit of the linear integrations along the x-axis is 1 and the lower limit is 0. To compute the x-direction centroidal coordinate as shown in Figure 11-5b, you then substitute into the formula and perform the integration as follows:

$$\overline{x} = \frac{\int_0^1 x\left(x - x^2\right)\cdot dx}{\int_0^1 \left(x - x^2\right)\cdot dx} = \frac{\frac{1}{3}x^3 - \frac{1}{4}x^4\Big|_0^1}{\frac{1}{2}x^2 - \frac{1}{3}x^3\Big|_0^1} = \frac{0.083}{0.167} = 0.500$$

Calculating Figure 11-5c's centroid location in the y-direction works much the same way as the x-direction calculation does, but with a few added issues. As in the x-direction, you start by modifying the incremental area calculation to become a linear integration calculation. The same dA expression you use for the x calculation still works. However, the y-distance requires a bit of work. In Figure 11-5c, notice that the variable y_i is actually different for each value of x. In this case, you need to transform that expression as well.

$$y = f_1(x) + \frac{1}{2}\left(f_2(x) - f_1(x)\right) = x^2 + \frac{1}{2}\left(x - x^2\right) = \frac{1}{2}\left(x^2 + x\right)$$

Remember, when you integrate with respect to a variable (in this case, dx), all variables in the expression must be in terms of x. In this example, the y location of the centroid is also a variable, but you know its relationship to the x variable because you know the equations of the upper and lower boundaries of the region.

Next, you substitute the expression for y into the centroidal equation:

$$\bar{y} = \frac{\int_0^1 \frac{1}{2}\left(x^2 + x\right)\left(x - x^2\right) \cdot dx}{\int_0^1 \left(x - x^2\right) \cdot dx} = \frac{\left[\frac{1}{6}x^3 - \frac{1}{10}x^5\right]\Big|_0^1}{\left[\frac{1}{2}x^2 - \frac{1}{3}x^3\right]\Big|_0^1} = \frac{0.067}{0.167} = 0.401$$

Notice here that you have to complete a multiplication of two polynomials before you can perform the integration. Now, if those polynomials are reasonably simple, this multiplication may not be that big an issue.

If you choose a slicing direction and the algebra and boundary-defining functions seem complex, try developing the incremental area calculations by slicing the region in the opposite direction as I describe in the nearby "Slicing a centroidal calculation differently" sidebar. After all, you still end up getting the same numeric results no matter how you slice it.

Taking advantage of symmetry

In some cases, you have to find the centroids of objects that have one or more *axes of symmetry*. An object is said to be symmetrical about an axis if the part of the object on one side of that axis is a *reflection,* or mirror image, of the part on the other side. Many shapes in engineering are symmetrical.

Some objects have one axis of symmetry, such as the images shown in Figure 11-6a and b. Other objects can have multiple axes of symmetry, such as the object in Figure 11-6c. And yet other objects can have an axis of symmetry that is neither horizontal nor vertical but rather *oblique* as in Figure 11-6d.

If you know that an object has an axis of symmetry, you can assume (and rightly so!) that the centroid location in the opposite direction must be located on that axis of symmetry. For example, Figure 11-6a has a vertical axis of symmetry. If you identify that this axis is located 5 millimeters to the right of the origin, you also know that the horizontal centroid distance is 5 millimeters to the right of the origin as well. You've just found one of the centroid locations without ever having to write a single equation!

Slicing a centroidal calculation differently

If the polynomials you need to multiply become lengthy, complex, or full of trigonometric functions, you may consider using a process similar to the following:

Sometimes you can simplify the math by changing the direction of the area slices. Reexamine the example in Figure 11-5 by using a horizontal incremental area as shown in the following figure.

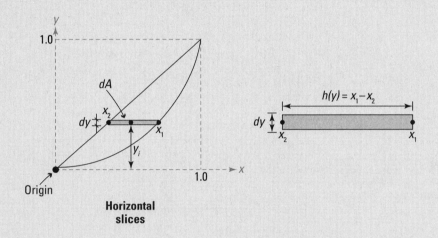

Horizontal slices

For example, if you repeat the calculation for the y-centroid coordinate but instead slice the incremental area dA horizontally, you get a different expression:

$$dA = (x_1 - x_2) \cdot dy = (\sqrt{y} - y) \cdot dy$$

For the horizontal slice, notice that the parameters x_1 and x_2 are actually dependent on the functions $f_2(x) = y = x_2$ and $f_1(x) = y = (x_1)^2$. Substituting into the same continuous centroid equation and evaluating the integral,

$$\bar{y} = \frac{\int_0^1 y(\sqrt{y} - y) \cdot dy}{\int_0^1 (\sqrt{y} - y) \cdot dy} = \frac{\left[\frac{2}{5}(y)^{\frac{5}{2}} - \frac{1}{3}y^3\right]\Big|_0^1}{\left[\frac{2}{3}(y)^{\frac{3}{2}} - \frac{1}{2}y^2\right]\Big|_0^1} = \frac{0.067}{0.167} = 0.401$$

Note that the result is identical to the traditionally calculated result in "Finding centroids of continuous regions." However, the polynomial multiplication for this method is a bit easier to work with. Unfortunately, depending on how you slice the continuous region (horizontal or vertical) to obtain your incremental area, you can either simplify your calculations or make them significantly more complicated. And you won't know which until you actually try.

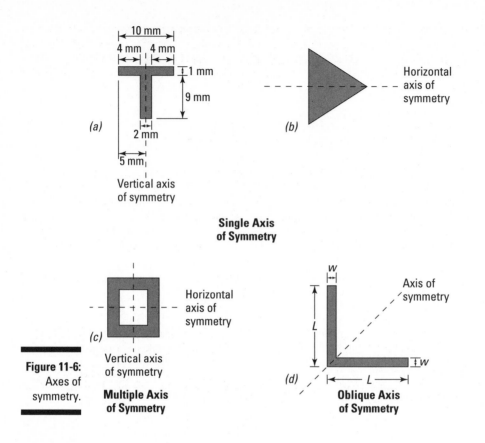

Figure 11-6: Axes of symmetry.

Understanding Centers of Mass and Gravity

Although you can use the calculations for centroids and centers of area in the previous sections with internal forces and external loads, self weight has its own special location requirements. To locate self weight, you first need to understand the difference between an object's center of mass and its center of gravity.

✔ **Center of mass:** An object's *center of mass* is the single location where its total mass can be applied as a single lumped value.

✔ **Center of gravity:** An object's *center of gravity* is the location on the object where resultant force due to gravity is acting. Self weight is a significant gravitational force on any object and is always located at the center of gravity.

Centers of mass and centers of gravity don't necessarily coincide with the centroids of geometric areas, although very often they do (check out "Getting to the Center of Centroids" earlier in the chapter for more on centroids). The center of gravity and center of mass also don't necessarily have to be contained within the boundary of the region.

Center of mass

The center of mass isn't necessarily tied to the geometric dimensions of the object but rather to the distribution of the mass within the object. For example, engineers often want a racing vehicle's center of mass as low as possible in order to ensure its stability at high speeds. However, the centroid of the vehicle is usually located at a position much higher up in the vehicle as a result of the physical dimensions of the automobile (see Figure 11-7).

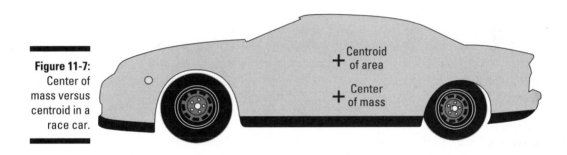

Figure 11-7:
Center of mass versus centroid in a race car.

For discrete regions, you can calculate the center of mass from the following expression:

$$x_{cm} = \frac{\sum_{i=1}^{n} x_i m_i}{\sum_{i=1}^{n} m_i} = \frac{x_1 m_1 + x_2 m_2}{m_1 + m_2}$$

For continuous regions, you have to fall back to the integral form (similar to the centroid calculations for continuous regions that I discuss earlier in the chapter), as follows.

$$x_{cm} = \frac{\int x \cdot dm}{m}$$

where m is the total mass of the object, and dm is the mass of an incremental slice of the object. The variable x represents the distance from the reference point to the center of the incremental mass, dm.

If you think this formula looks very similar to the centroid equations I discuss earlier in the chapter, you're right. The only difference is that instead of using the region's geometric area in your calculation, you're now using its mass. Consider the system of two masses (m_1 and m_2) shown in Figure 11-8.

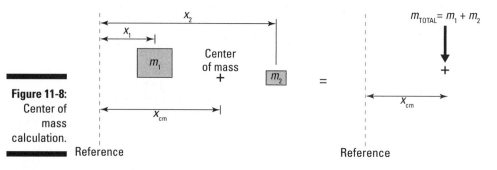

The center of mass is directly related to the location of the mass of each sub-region and its position relative to a reference location, so always include a reference or origin when measuring the center of mass.

Center of gravity

As long as the gravitational pull on an object is uniform, the centers of gravity and mass share the same position. On Earth, you can reasonably assume this scenario to be the case because the Earth's gravitational field typically doesn't fluctuate very much over short distances.

In this book, I assume that the gravitational field is constant for the object, and thus that the center of mass and the center of gravity occur at the same location.

Chapter 12

Special Occasions in the Life of a Force Vector: Moments and Couples

*I*n earlier chapters, I discuss the concept of forces and explain how a force (such as a bat) pushing on an object (such as a baseball) causes that object to move in the direction of the applied force (known as *translating*). However, not all actions cause an object to move or translate; some cause rotation.

In this chapter, I illustrate the rotational behaviors, including moments, couples, and concentrated moments, of objects and present the equations that let you calculate the behaviors that cause these rotational effects.

I Need a Moment: Exploring Rotation and Moments of Force

Think about a pinwheel. By blowing on the pinwheel, you're actually applying a force to it. Unlike the examples in previous chapters, the pinwheel doesn't move in the direction of the force because it's tied to a stick. But it does spin or *rotate*. The pinwheel stays in place, but it spins. The harder you blow (the more force you apply) to a pinwheel, the faster it spins.

Rotational behaviors can also occur in objects that are translating. Rolling is a combination of both linear motion (or translation) and rotation. The old tumblin' tumbleweeds that you see in classic Westerns are a great example. These dried plants move along the dusty countryside (translation), turning under the force of the wind (rotation).

The general physics definition of a *moment* is always "force times distance." This simple concept doesn't account for several other considerations that come into play, but it's efficient enough to illustrate the units of a moment: Newton-meters (N-m) for SI units and pound-feet (lb-ft) for U.S. customary units.

Because a moment is often a direct result of the action of a force vector, a moment is also a vector and must also have similar characteristics. Like any vector, a moment vector has a magnitude, point of application, and sense (see Chapter 4 for more on these basic characteristics).

The *magnitude* of the moment is a measure of the intensity of the rotational effect. Instead of having a unique *point of application* (location in space) or a defined *line of action* (line in space on which the vector is acting) as forces do, the moment actually rotates around an axis called the *axis of rotation.* The *sense* is the direction of rotation about its axis of rotation. You usually represent it as a clockwise (negative) or counterclockwise (positive) behavior. An axis of rotation can either be within the object, which results in a rotational behavior known as *pivoting,* or outside an object, which results in a rotational behavior known as *orbiting.*

Figure 12-1 shows you the similarities between force vectors and moment vectors. You display a force with a single-headed arrow, as I describe in Chapter 4. The figure also shows you two ways for depicting a rotation — one with a circular arrow, and another with a double-headed arrow. See the similarities between moment and force sketches? Don't worry, I discuss these methods for drawing moments later in this chapter.

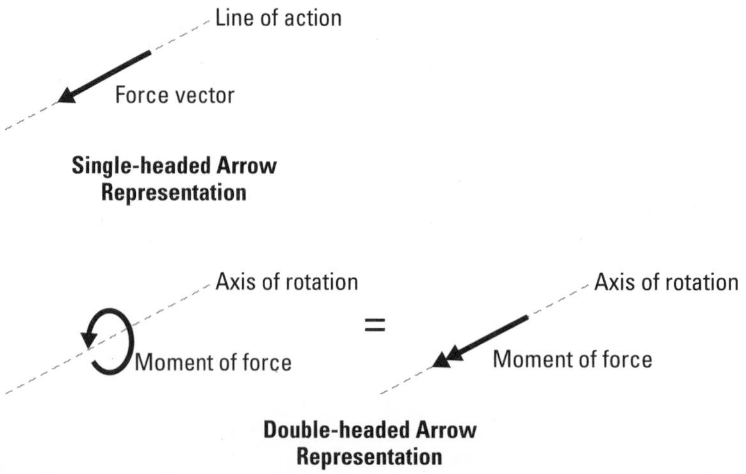

Line of action

Force vector

Single-headed Arrow Representation

Figure 12-1:
Rotation about an axis.

Axis of rotation

Moment of force

=

Axis of rotation

Moment of force

Double-headed Arrow Representation

Just as with the line of action for a force, the axis of rotation for a moment doesn't have to be aligned with a Cartesian axis. (Chapter 5 gives you the low-down on Cartesian axes.) An axis of rotation can have any unique orientation in space. In fact, one of the difficulties you experience whenever you create a moment vector is actually defining the direction of the axis of rotation. I show you how to do just that in the "Using unit vectors to create moment vectors" section later in this chapter.

Developing rotational behaviors: Meeting couples and concentrated moments

You can develop a rotational behavior in several different ways. Some require forces, some require distances, and some require neither of those. In statics, rotational behaviors are created by one of three principal methods: one force and a distance, two parallel forces separated by a distance (or couples), and a concentrated rotational effect (or concentrated moment), which I cover in the following sections.

One force and one distance

Consider the behavior of an open door when you push on it as shown in Figure 12-2.

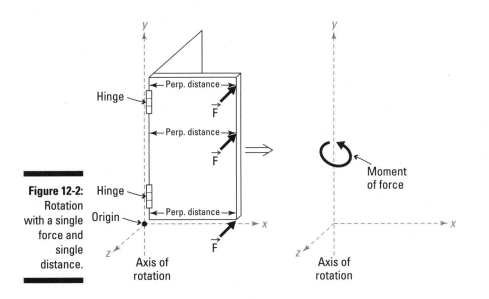

Figure 12-2: Rotation with a single force and single distance.

In a door assembly, doors typically hang from several hinges that are aligned along a single vertical axis of rotation. When you push on the door (apply a force), the resulting action is that the door begins to move. Regardless of where you apply the force on the door (at the top, on the handle, or by your foot on the bottom of the door), the same resulting behavior (a moment) occurs along the axis of rotation. This moment is what creates the rotation that results in the door swinging open or closed.

Two parallel forces and a distance: Couples

A *couple* is a type of moment produced by two parallel forces of the same magnitude acting in opposite directions and separated by a distance that result in a rotational behavior on the object. One example of a couple is the forces from your fingers on a doorknob. As you turn the knob, one finger is pushing up on the side of the doorknob and the other is pushing down. These two forces together cause the knob to rotate. Actually opening or closing the door requires that you push on the door after you have turned the knob, which is the scenario that I discuss the preceding section. Opening a door actually requires two moments (mechanically speaking, that is).

As another example, imagine you're driving your car with your hands at "ten and two" in proper driving fashion. As you're traveling down the road, a ball bounces out into the lane in front of you. Out of instinct, you quickly turn the wheel clockwise to swerve around the obstacle by pushing up on the left edge of the wheel with your left hand while pulling down on the right edge of the wheel with your right hand as shown in Figure 12-3.

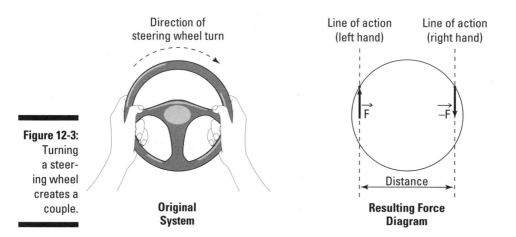

Figure 12-3: Turning a steering wheel creates a couple.

The behavior of your hands on the wheel when you turn clockwise is actually caused by two separate forces applied to the wheel. The force on the left side of the steering wheel is acting upward, while the force on the right side of the wheel is acting downward. In this example, the resultant of these two forces is zero ($\mathbf{F} + (-\mathbf{F}) = 0$) or balanced. The wheel isn't actually translating (or moving) in any direction, but it still experiences a rotational behavior — a couple, from these two forces despite having no net force (or zero *resultant forces*) acting on it.

No distance? Concentrated moments

Another scenario that can cause an object to rotate is the application of a force or moment to another connected object. This resulting and transmitted moment is known as a *concentrated moment*. In fact, these mechanisms are very common in mechanical shaft design. The purpose of a shaft is to transmit a force or moment from one location in an object or mechanism to another through the action of the shaft. Consider Figure 12-4, which shows a single force F acting on an L-shaped bar.

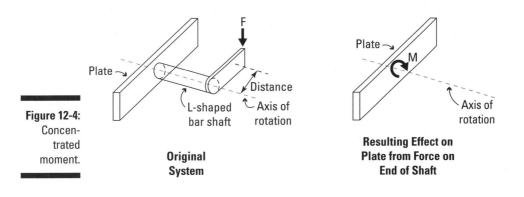

Figure 12-4: Concentrated moment.

Original System

Resulting Effect on Plate from Force on End of Shaft

In this example, the force causes the shaft to rotate about its axis of rotation, which in turn creates a resulting moment on the plate at the location where the shaft and plate are connected.

This *transmission of moments* can occur through multiple objects as well. An example of this type of situation occurs in the drive train of your automobile. The engine of your car causes the transmission to rotate, which in turn causes the axles to rotate, which in turn cause the wheels to rotate. Now obviously, moving a car down the road requires many more factors, but the overall concept of the transmitted moment is still valid.

Taking on torque and bending: Types of concentrated moments

You can create different types of rotational effects depending on which axis a concentrated moment is causing a rotation about. In statics, two of the most common effects are bending and torque (or torsion), shown in Figure 12-5.

✔ **Torque:** A *torque* is a *torsional moment,* or one that causes rotation about a longitudinal axis of an object that causes a twisting action.

✔ **Bending:** A *bending moment* is a moment that is applied about an axis that is perpendicular to a member's longitudinal axis, or applied in the plane of the *cross section* (or a slice through the member — see the shaded region in Figure 12-5). That is, if it isn't a torsional moment, it has to be a bending moment.

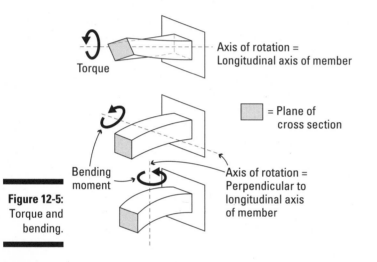

Figure 12-5: Torque and bending.

Because moments are also vectors, a resultant moment may have components that produce torque and multiple bending effects simultaneously. I talk more about computing these effects in Chapter 20.

Getting a handle on the right-hand rule for moments of force

In statics, you can find several different variations of right-hand rules that prove to be very useful as you start working problems. One of the most useful versions helps you determine the sense of a moment about its axis of rotation.

You can determine the sign (or sense) for moments by making an L-shape with your right thumb and forefinger. Align your thumb with the positive *x*-axis of your Cartesian coordinate system and then line up your forefinger with the positive *y*-axis at the same time. (This orientation may feel a bit awkward at first, but it does work!). Bend your middle finger naturally so that it's perpendicular to your thumb and forefinger and pointing outward from your palm. Your middle finger represents the direction of the positive *z*-axis (see Figure 12-6).

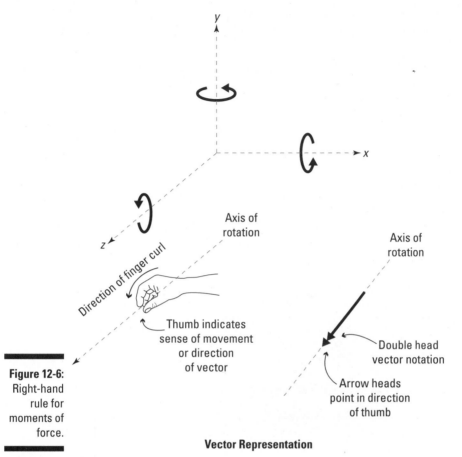

Vector Representation

Figure 12-6:
Right-hand
rule for
moments of
force.

WARNING!

The right-hand rule only works with the right hand! If you use your left hand by mistake, the direction of your z-axis will be backward.

After aligning your fingers with the Cartesian system, if your moment is about one of the Cartesian x-, y-, or z-axes, you can determine the sense of the moment by looking at the end of the finger that is parallel to the axis of rotation of the moment. In three dimensions, a moment is positive about the x-axis if it's acting counterclockwise when you're looking at the tip of your thumb. The same applies to the y-axis and your forefinger and to the z-axis and your middle finger.

Calculating a Moment with Scalar Data

To calculate the magnitude of a moment of a force, you need to include two pieces of information in your computations: the force and the distance. In general, you can calculate the magnitude of the moment from the following equation:

$$\left|\overline{M}\right| = \left|\overline{F}\right| \cdot \left|\overline{d}\right| = \text{force} \cdot \text{distance}$$

This equation represents the scalar form of the moment calculation. When you use this formula, you're actually only calculating the magnitude of the moment — you haven't actually considered the sense of the line of action required to fully define it as a vector, nor have you defined the axis of rotation. In the scalar equation, the distance term is the distance from the axis of rotation.

When using the scalar calculation, the distance from the axis of rotation to the line of action of the force must be perpendicular. No exceptions!

In two dimensions, the x and y Cartesian axes are usually in the plane of the page, which results in the third axis, the z-axis, being out of the page because it must be perpendicular to the two-dimensional axes. A two-dimensional moment of a force located in the xy Cartesian plane is always about the z-axis (the axis of rotation is parallel to the z-axis). In vector terms, this fact means that the moment in the xy plane has a unit vector direction of **k** (either positive or negative depending on the sense of the moment).

Planar rotation about a point

In this section, I show you how to perform the calculation of the moment after you know the location of the point in space. I actually explain how to choose the necessary moment locations for your calculations when I discuss the various techniques of Part VI.

The major drawback of the scalar method of computing moments is that you have to assign the sense of the vector based on logic.

Suppose you want to calculate the moment of the force in Figure 12-7a (which shows a force with a magnitude of 300 pounds acting at an angle of 60 degrees above the negative x-axis) about Point A. Just follow these steps:

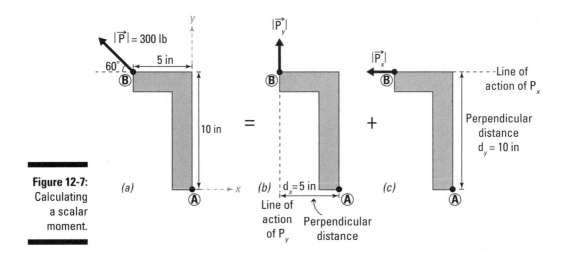

Figure 12-7:
Calculating
a scalar
moment.

1. **Break the vector into components in the *x*- and *y*-direction.**

   The first step is to compute the *x*- and *y*-components by using the basic trigonometry principles that I discuss in Chapter 8.

   $$\left|\overrightarrow{P_x}\right| = 300 \cdot \cos(60) = 150 \text{ lb } \leftarrow$$

   $$\left|\overrightarrow{P_y}\right| = 300 \cdot \sin(60) = 259.81 \text{ lb } \uparrow$$

   After you've computed the components, you can apply them to the original object one at a time, as shown in Figures 12-7b and 12-7c.

2. **Calculate the moment contribution of each component that you calculated in Step 1 about the point of interest.**

   For the vertical force **P**$_y$ shown in Figure 12-7b, you can calculate the moment by multiplying by the perpendicular distance.

   If the dimensions of the force measured to the point are in directions parallel to the Cartesian *x*- and *y*-axes, you want to break the force into components parallel to those axes as well.

   In this case, because the force **P**$_y$ of Figure 12-7b is vertical, you need to use the horizontal distance when you calculate the moment at Point A (M_{A1}).

   $$\left|\overrightarrow{M_{A1}}\right| = \left|\overrightarrow{P_y}\right| \cdot \left|\overrightarrow{d_x}\right| = (259.81 \text{ lb}) \cdot (5 \text{ in}) = 1{,}299 \text{ lb} \cdot \text{in (clockwise)} = -1{,}299 \text{ lb} \cdot \text{in}$$

   The direction *clockwise* is determined by considering which direction the force would rotate the object if A were pinned in its current position. This clockwise statement is an indication of the sense of the vector. As I discuss in "Getting a handle on the right-hand rule for moments of force" earlier in the chapter, a moment about the *z*-axis is considered positive if it's acting counterclockwise about the axis of rotation; therefore, the clockwise moment of the force is actually a negative moment.

Similarly, you can calculate the moment for the horizontal force $\mathbf{P}_x$ of Figure 12-7c. In this case, you need to use the vertical distance, which is perpendicular to the line of action of the horizontal force.

$$\left|\overrightarrow{M_{A2}}\right| = \left|\overrightarrow{P_x}\right| \cdot \left|\overrightarrow{d_y}\right| \quad = (150 \text{ lb}) \cdot (10 \text{ in}) =$$
$$= 1{,}500 \text{ lb} \cdot \text{in} \, (\text{counterclockwise}) = +1{,}500 \text{ lb} \cdot \text{in}$$

3. **Compute the net effect of the moments of the component forces about the location of interest.**

$$\left|\overrightarrow{M_A}\right| = \left|\overrightarrow{M_{A1}}\right| + \left|\overrightarrow{M_{A2}}\right| = -1{,}299 + (+1{,}500) = +201 \text{ lb} \cdot \text{in}$$
$$= 201 \text{ lb} \cdot \text{in} \, (\text{counterclockwise})$$

Because the net magnitude is positive, you know that the net moment about Point A is acting in a positive (or counterclockwise) direction with a magnitude of 201 lb-in.

Determining the magnitude and sense of a two-dimensional couple

You can treat a couple as either two separate forces or as a pair of forces separated by a distance. Both come out to the same magnitude value. You compute the moment couple of a pair of forces by relying on the same general principle of the force times distance relationship. Consider the couple shown in Figure 12-8; it's created by a pair of 200-Newton forces separated by a distance of 2 meters.

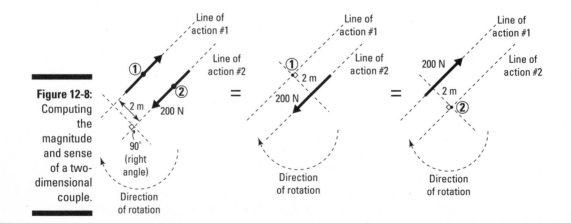

Figure 12-8: Computing the magnitude and sense of a two-dimensional couple.

To calculate the magnitude of the force couple, you use the formula from the preceding section:

$$\left|\overline{M}\right| = (\text{force}) \cdot (\text{distance}) = (200\text{ N}) \cdot (2\text{ m}) = 400\text{ N} \cdot \text{m}$$

where the force is the magnitude of one of the forces in the couple and the distance is the perpendicular distance between the lines of action of the forces. To determine the sense of the couple, you need to choose both a point on one of the forces' lines of action as well as the other force itself. For this example, if you choose Point 1 on the line of action #1 as your reference point, you'd choose the force on the line of action #2. Now, to determine sense of the moment, you consider the direction of rotation of your selected force about your selected point. In this case, the selected force wants to rotate about Point 1 in a clockwise direction (which indicates a negative sense), so you can say the sense of this couple is negative. Your final solution may look something like the following:

$$\left|\overline{M}\right| = -400\text{ N} \cdot \text{m} = 400\text{ N} \cdot \text{m (clockwise)}$$

Notice that if you choose the other point (Point 2) as your reference point and the force on the line of action #1, you end up with the same rotational sense.

Calculating a Moment by Using Vector Information

Moment magnitudes are pretty simple to calculate by using scalar information (and the preceding section), and a little logic helps you determine the sense, particularly when you can compute (or already know) the required perpendicular distance. For two-dimensional problems, determining a perpendicular distance may not be all that difficult, especially if the axis of rotation is aligned to one of the Cartesian axes.

But what happens when you have a three-dimensional force that's creating a moment about some random point in space? Finding that perpendicular distance can be a little rough, especially when the axis of rotation isn't aligned conveniently. You almost always have to fall back to using the following vector form to compute the moment vector **M.**

$$\overline{M} = \vec{r} \times \vec{F}$$

where **r** is a position vector from the axis of rotation to any point on the line of action of the force vector **F.** *Warning:* The strange $\times$ in the equation isn't the multiplication symbol but rather is called the *cross product,* an operation performed on two different vectors that produces a third vector that's perpendicular to each of the original vectors. You read this equation as "r cross F." I show you how to actually perform this calculation a little later in this section.

The major advantage of using the vector form over the scalar form when calculating a moment is that you don't have to worry about calculating those pesky perpendicular distances because they're already handled by the vector mathematics contained within the cross product calculation. In fact, the vector solution forms will always work, although for two-dimensional problems, the scalar math calculations are often a lot simpler than performing a cross product.

If you align your coordinate system such that one of the Cartesian axes is parallel to the axis of rotation, you may find that the notation of your moment vector is significantly simpler.

Completing the cross product

The most difficult part of creating a moment vector is actually the computation of the cross product. Although it's not mathematically difficult, it can be a somewhat lengthy process (as you can see in the equation I show you in Chapter 6). Figure 12-9 shows a three-dimensional vector and a center of rotation at Point 1.

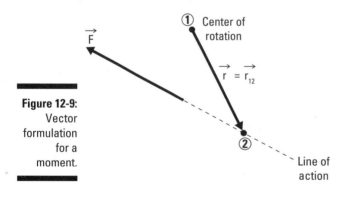

Figure 12-9:
Vector
formulation
for a
moment.

To compute the moment, you need two pieces of information. The first is the Cartesian vector formulation for the force that is creating the moment. You can use any of the techniques that I describe in Part I to help you create the force vector. The second piece of information you need is the position vector that starts at Point 1 at the center of rotation (or point of interest) in Figure 12-9 and connects to Point 2, which is a point at any location along the line of action of the force. It doesn't matter where you place the second point as long as it's somewhere along the force's line of action. Normally, you pick that second point as the point of application of the force (because you often know those coordinates) or a place on the geometry where the dimensions are already defined for you.

After you have this information, you can substitute the scalar component values into the cross product formula to compute the magnitude of the moment. As I cover in Chapter 6, one technique for solving a cross product calculation is using a *determinant*. The determinant form is shown in the following equation.

$$\overline{M} = \vec{r} \times \vec{F} = \begin{vmatrix} \hat{i} & \hat{j} & \hat{k} \\ r_x & r_y & r_z \\ F_x & F_y & F_z \end{vmatrix} \quad \begin{matrix} \text{line 1} \\ \text{line 2} \\ \text{line 3} \end{matrix}$$

In the vector formulation, you include the unit vectors for the Cartesian axes on line 1. The position vector information goes on line 2, where you include the scalar magnitudes of the components of the position vector. On line 3, you input the force vector information, which includes the component magnitudes of the force. For lines 2 and 3, you place the x-component information for both the position and force vectors in the column below the x-direction unit vector ($\mathbf{i}$), the y-component information below the y-direction unit vector ($\mathbf{j}$), and the z-component information below the z-direction unit vector ($\mathbf{k}$). If your position vector or force vector doesn't have a particular value for an x-, y-, or z-component, you simply put a zero value in that location. The final answer from this calculation is a vector representation of the moment of the force about the center of rotation.

Using unit vectors to create moment vectors

In some cases, you may know the magnitude of an applied moment about an axis of rotation, particularly if you've used scalar computations to compute the moment. You can create a unit vector defining the direction of the moment by creating a different unit vector that describes the direction of the axis of rotation.

You can always relate any vector to its magnitude and direction. For moments, you use something like the following:

$$\overline{M} = \left| \overline{M} \right| \cdot \overrightarrow{u_M}$$

In this equation, $\mathbf{u_M}$ is a unit vector in the direction of the sense of the moment (or the direction of your thumb if you've used the right-hand rule for moments, discussed earlier in the chapter).

Rule of Sarrus

A fairly easy shortcut known as the rule of Sarrus can actually eliminate some of the troubles with signs that may pop up when you compute the determinant in your moment vector calculations. The following procedure simplifies the computation:

1. **Enter the values into the determinant as described in the nearby section "Completing the cross product."**

2. **Augment the matrix from Step 1 by copying the i and j columns to the positions shown in the following equation.**

   Your calculation now looks like this:

$$\vec{M} = \vec{r} \times \vec{F} = \begin{vmatrix} \hat{i} & \hat{j} & \hat{k} \\ r_x & r_y & r_z \\ F_x & F_y & F_z \end{vmatrix} \begin{matrix} \hat{i} & \hat{j} \\ r_x & r_y \\ F_x & F_y \end{matrix}$$

3. **Starting at the upper left i value, multiply everything on the diagonal from upper left to lower right; repeat this process for both the j and k values and then add these three multiples.**

   You get something like the following expression:

$$\hat{i} \cdot r_y \cdot F_z + \hat{j} \cdot r_z \cdot F_x + \hat{k} \cdot r_x \cdot F_y$$

4. **Starting at the lower left F_x value, multiply everything on the diagonal from lower left to upper right; repeat this process for both the F_y and F_z values and then add these three multiples.**

   Here's what that expression looks like:

$$\vec{M} = \vec{r} \times \vec{F} = \begin{vmatrix} \hat{i} & \hat{j} & \hat{k} \\ r_x & r_y & r_z \\ F_x & F_y & F_z \end{vmatrix} \begin{matrix} \hat{i} & \hat{j} \\ r_x & r_y \\ F_x & F_y \end{matrix}$$

$$\hat{k} \cdot r_y \cdot F_x + \hat{i} \cdot r_z \cdot F_y + \hat{j} \cdot r_x \cdot F_z$$

5. **Subtract the results of Step 4 from the results of Step 3.**

   This value is the vector formulation of the moment and the mathematical solution of your determinant.

$$\vec{M} = \underbrace{\left[\hat{i} \cdot r_y \cdot F_z + \hat{j} \cdot r_z \cdot F_x + \hat{k} \cdot r_x \cdot F_y \right]}_{\text{Step 3}} - \underbrace{\left[\hat{k} \cdot r_y \cdot F_x + \hat{i} \cdot r_z \cdot F_y + \hat{j} \cdot r_x \cdot F_z \right]}_{\text{Step 4}}$$

The results of this calculation are identical to those from the cross product methodology I describe in Chapter 6.

Consider the example shown in Figure 12-10.

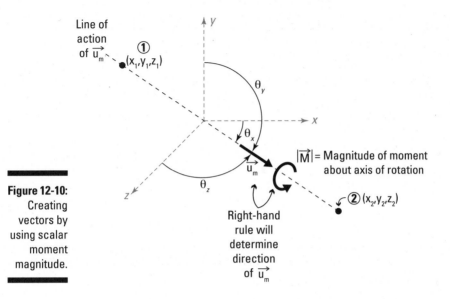

The unit vector describing the sense and direction of the moment is actually the same unit vector that describes the direction of the axis of rotation.

In Chapter 5, I show you three different techniques for creating unit vectors for forces; if you can create a unit vector describing a force's line of action, you can use the same techniques to define an axis of rotation, as the following list shows.

✔ **Using position vectors:** If you know two points on the axis of rotation, you can use the position vector method to create a unit vector by dividing a position vector by its magnitude. In Figure 12-10, the position vector goes from Point 1 to Point 2 (both of which lie on the axis of rotation) because it must be in the same direction as the unit vector $\mathbf{u_M}$. The denominator of this equation is just the magnitude of the position vector in equation form.

$$\overrightarrow{u_{12}} = \frac{(x_2 - x_1)\hat{i} + (y_2 - y_1)\hat{j} + (z_2 - z_1)\hat{k}}{\sqrt{(x_2 - x_1)^2 + (y_2 - y_1)^2 + (z_2 - z_1)^2}} = \overrightarrow{u_M}$$

✔ **Using direction cosines:** If you happen to know the angles between each of the Cartesian axes and the axis of rotation, you can use the direction cosine formulation (which follows) to create your unit vector for the moment.

$$\vec{u}_{AXIS} = \cos(\theta_x)\hat{i} + \cos(\theta_y)\hat{j} + \cos(\theta_z)\hat{k} = \overrightarrow{u_M}$$

Using Double-Headed Arrows to Find Moment Resultants and Components

In Chapter 4, I show you how to draw a double-headed vector. These two-headed monsters are actually extremely useful when you start working with moments. One of the major points of confusion with moments has to do with the concept of rotation. How do you accurately depict a rotation behavior about a point or axis?

In two dimensions, you can easily illustrate a circular arrow depicting the direction of the moment because the circular arrow almost always acts around the z-axis. However, three-dimensional cases, where the moment can act around any axis in space, are a little harder to illustrate. And when you're computing components from the rotational depiction, such illustration becomes next to impossible.

For this reason, I like to use the double-headed vector notation as shown in Figure 12-11 (though I only use it when I'm working with moments) which lets you handle moments easily and effectively in the same manner as you would treat a force vector.

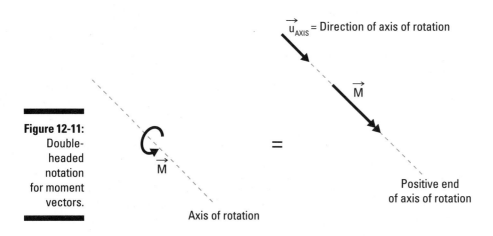

$\vec{u}_{AXIS}$ = Direction of axis of rotation

$\vec{M}$

Figure 12-11:
Double-headed notation for moment vectors.

$\vec{M}$

=

Positive end of axis of rotation

Axis of rotation

I start by using the right-hand rule for moments to determine the sense and the direction of the unit vector to help me define the axis of rotation. (See "Getting a handle on the right-hand rule for moments of force" earlier in the chapter.) Under the right-hand rule, the double-headed vector points in the same direction as your thumb. In Figure 12-11, the moment produces a counterclockwise moment about the axis of rotation, which means the double-headed vector must point toward the positive end of the axis.

Although this graphical change may seem a little pointless at first, this transformation actually allows you to utilize the same vector component and resultant manipulations you use on force vectors in Chapters 7 and 8. Figure 12-12 illustrates the similarities between the calculations of single-headed notation and those of the double-headed notation.

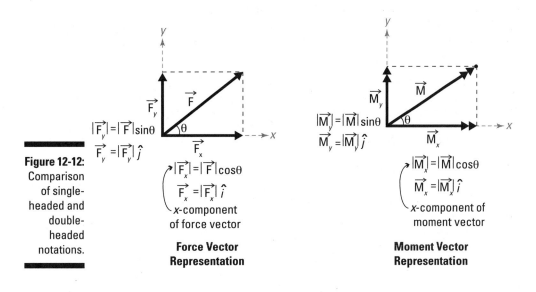

Figure 12-12: Comparison of single-headed and double-headed notations.

Force Vector Representation

Moment Vector Representation

In this figure, the components for both notations are computed using the exact same vector formulations. For force vectors, the force component in the x-direction is the portion of the force vector acting in the x-direction. For moment vectors, the component in the x-direction is actually the portion of the moment that is acting about the x-axis. That is, the unit vector defining the direction of the moment's double-headed arrow representation is actually acting in the x-direction, but the final behavior is acting around the axis.

For resultants, the same methodology applies. Figure 12-12 shows that you can create a force vector for this two-dimensional application by simply adding the vector components of the single-headed notation in the x- and y-directions:

$$\overrightarrow{F_{RES}} = \sum \overrightarrow{F} = \overrightarrow{F_x} + \overrightarrow{F_y}$$

Likewise, for the moments, you can create a resultant moment by simply adding the vector components of the double-headed notation about the x- and y-axis:

$$\overrightarrow{M_{RES}} = \sum \overrightarrow{M} = \overrightarrow{M_x} + \overrightarrow{M_y}$$

Relocating a Force by Using a Moment: Equivalent Force Couples

When performing your basic statics calculations, you often find relocating a force from one point to another convenient. By creating an *equivalent force couple,* you can move a force vector to a new location by simply relocating the force and creating a new moment at the new location.

An *equivalent system* is two systems that experience both the same translational and rotational behaviors.

Figure 12-13 shows a rigid body with two different points, A and B. In the first picture, a force vector **F** acts at Point A. Notice that the force **F** at A is *eccentric to* (or not acting through) Point B.

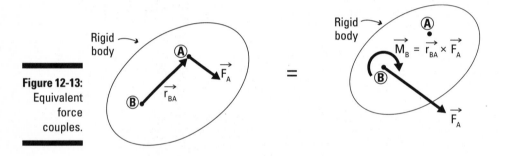

Figure 12-13: Equivalent force couples.

To produce the same translational effect on this *rigid body* (a body not deformed by the force), you simply need to relocate the force at Point A to its new position at Point B. However, after you move the force from Point A to Point B, the rotational behavior of the object changes. In order to capture the rotational effects of the force at Point A with respect to Point B, you have to include an additional rotational effect, or a moment, which you can compute with the following formula:

$$\overrightarrow{M_B} = \overrightarrow{r_{BA}} \times \overrightarrow{F_A}$$

where $\mathbf{r_{BA}}$ is a position vector from the new point (B) to the original point (A). The methods for computing the position vectors and force vectors remain unchanged. I show you more about the uses and implementation of this idea in Part VI.

Part IV

A Picture Is Worth a Thousand Words (Or At Least a Few Equations): Free-Body Diagrams

The 5th Wave   By Rich Tennant

STATICS MEN'S CLUB

Let's hear it for Amber, who's gonna show us what she knows aboooouut three dimensional pinned supports!

In this part . . .

Free-body diagrams (F.B.Ds) are a vital part of solving a statics problem, so these chapters give you a basic checklist for constructing the pictures that describe an object and the loads acting on it. You also discover how to include external applied loads, internal forces, and self weight. I introduce the different types of physical restraints (known as *support reactions*) and show you how to include them on your F.B.D. I then illustrate how to simplify complex free-body diagrams by moving loads from one point on an object to another while maintaining the same object's behavior.

Chapter 13

Anatomy of a Free-Body Diagram

- -

In This Chapter

▶ Knowing what to include on a free-body diagram

▶ Including external force effects

▶ Applying internal force effects

▶ Restraining motion and rotation with support reactions

▶ Remembering self weight

- -

Ask photographers and artists about the pictures they've created, and you inherently hear about the emotions and feelings that they were trying to capture as they portrayed the physical object of their work. Pictures in statics provide a different purpose — something a bit more unemotional and unbiased; after all, statics is a science. However, the facts do show that a picture can serve as a very handy and even necessary tool; these pictures are what allow you to create those (objectively) super-awesome equations of equilibrium. Without a properly detailed picture (known as a free-body diagram), the game is over before you even get off the bench.

In this chapter, I describe the four types of forces that must be included on a free-body diagram and discuss the proper technique for displaying them.

Free-Body Diagrams in a Nutshell

The picture that you draw in statics is known as the *free-body diagram* (or F.B.D. for short) and represents the physical condition of the rigid object you want to analyze, including dimension data and the forces acting on the system. Free-body diagrams can be complex pictures of multiple objects and systems, or diagrams of a smaller subcomponent of a larger piece within a system. Each representation must still obey all laws of physics. Without a proper F.B.D. sketch, correctly analyzing a problem in any field of engineering and physics is extremely difficult, if not impossible.

I've always found a checklist useful for remembering what to include on any given free-body diagram.

You can classify the majority of forces on an object into four separate categories, each of which becomes an item on the checklist: external forces, internal forces, support reactions, and self weight. Just remember, you need to include the forces themselves as well as all information that locates their *point of application* (physical location on the object or in space where the vector is acting — see Chapter 4). Check out Chapter 9 for more on how all of these categories affect concentrated loads; Chapter 10 discusses how external and internal forces and self weight relate to distributed loads.

- ✔ **External forces:** *External forces* are the forces exerted on a *rigid body* (which isn't deformed by the force) by sources outside the body. A ball thrown at a wall exerts an external concentrated force at the point of impact; the weight of snow on your roof exerts a vertical distributed external force on the roof.

- ✔ **Internal forces:** *Internal forces* are the forces exerted within a rigid body. The tension in a rope and the compressive force in the leg of the chair you're sitting on are both examples of internal forces.

- ✔ **Support reactions:** *Support reactions* are the physical restraints, such as door hinges and bridge piers, that prevent a rigid body from moving.

- ✔ **Self weight:** *Self weight* (in both concentrated and distributed form) is the force due to gravitational effects on the mass of the object.

In addition to these four categories of forces, you also need to include the necessary dimensions and angles that help you properly define their lines of action and points of application. I explain more in the coming sections about how you actually draw each on the free-body diagram.

Displaying External Forces

External forces are typically the easiest forces for you to determine because they're often the result of a measureable action — you typically know the sizes and shapes of their distributions. These forces include both *concentrated forces* (or *point loads*) and *distributed forces* (forces over area), as I discuss in Chapters 9 and 10, as well as *concentrated moments,* which I discuss in Chapter 12.

Consider the example drawing in Figure 13-1, which shows a man pushing horizontally on a crate with a force of 100 Newton. The crate also has a very heavy lid of uniform thickness resting on its top.

If you want to draw the external forces acting on just the crate (without the lid) you have to apply a bit of logic and reasoning to determine the sources of external forces. Figure 13-1 has two external forces — concentrated and distributed — acting on the rigid crate. The following sections show you how to depict those forces in your F.B.D.

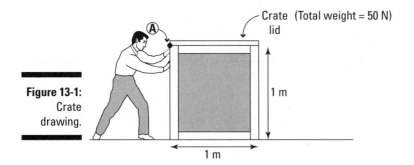

Figure 13-1:
Crate
drawing.

Crate (Total weight = 50 N)
lid

1 m

1 m

Portraying concentrated forces

Concentrated forces are typically the easiest to portray on an F.B.D. For
any concentrated force vector, you know the magnitude, sense, and point
of application (or in some cases, the line of action). If you know the point of
application, you can use that point directly on the free-body diagram. If you
only know the line of action, you need to locate the line on the F.B.D. and
then apply the force somewhere along that line.

The first external load acting on the crate in Figure 13-1 is from the force
exerted by the man as he pushes horizontally at Point A. The force of the
man's hands on the edge of the crate in this situation is depicted as a concentrated load because the force is applied at a single point. Figure 13-2 shows
how you can represent this force as a concentrated load.

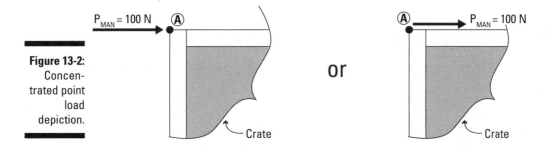

Figure 13-2:
Concen-
trated point
load
depiction.

$P_{MAN} = 100$ N (A)

or

(A) $P_{MAN} = 100$ N

Crate

Crate

To properly represent a concentrated force, you must include information
about the three basic properties of a vector: magnitude, sense, and point of
application, all of which I cover in Chapter 4.

✔ **Magnitude:** The *magnitude* is the vector's length. In both parts of Figure 13-2, notice that the magnitude of the force is given as 100 Newton acting horizontally. The 100 Newton in this example is actually the magnitude of the vector you're drawing.

✔ **Sense:** The *sense* is the direction in which the vector is acting, which you can help determine by figuring out which direction the object would want to move as a result of the force. Put yourself into Figure 13-2; the crate would want to slide to the right if you pushed hard enough. Therefore, you can reason that the vector's sense is also to the right.

✔ **Point of application:** The point of application of this force vector is situated at Point A because that's where the man's hands are located. The most conventional means of representing this vector is to apply the tail of the vector at the point of application as shown in Figure 13-2b. However, when forces are pushing on objects, placing the head of the vector at the point of application instead (as shown in Figure 13-2a) can be a more convenient reminder. The principle of transmissibility (covered in Chapter 9) tells you that the two drawings shown in Figure 13-2 behave identically as long as the crate is considered to be perfectly rigid.

Depicting distributed forces

You draw distributed loads similar to their concentrated counterparts — they have a sense to define their direction, and a magnitude (defined as its intensity). However, a distributed load has no specific point of application because the load is spread out over a line or region. You must include dimensional information that shows where the distributed load begins as well as where it ends.

The second type of external load on the crate in Figure 13-1 is from the weight of lid as it sits on the crate. When you look at just the lid, this 50-Newton weight is actually the self weight of the lid itself. However, because your F.B.D. is of the forces on just the crate itself, this force becomes an external force on the crate.

Don't be too alarmed about the difference between self weight and external forces. As long as you include the force at its proper location on the object, the solution process is identical. Just remember that self weight can be either a single value *(lumped mass)* or a distributed load over a length or area *(continuum)*. For more on self weight, flip to Chapters 9 and 10.

If you consider the weight of the lid as a uniformly distributed load, you can compute that this 50-Newton external force is acting over a length of 1 meter (the dimension of the lid). Thus, the external distributed load from the lid is computed as $w = \dfrac{50\text{ N}}{1\text{ m}} = 50\,\text{N/m}$, acting downwards.

To display this distributed force on the F.B.D., you need the same three basic requirements (magnitude, sense, and direction) as for the concentrated loads in the preceding section:

- ✔ **Magnitude:** As for the concentrated loads, the magnitude (or intensity) of the load distribution is 50 Newton per meter.

- ✔ **Sense:** Because the load is actually coming from a self weight, the sense of the distributed load acts in the direction of gravity. I'm assuming this object is on planet Earth (because scientists haven't found any evidence of crates on Mars, at least not yet), so gravity and therefore the sense are acting downward.

- ✔ **Point of application:** The point of application of this load depends on whether the load is distributed or concentrated. If the load were concentrated, it would act at the center of mass of the lid (in this case, the midpoint). But, because you've calculated the weight of the lid as a distributed load and you know that the lid's thickness is uniform (or constant), this load is evenly spread (or uniformly distributed) over the entire length.

In addition to these three requirements, you also need a couple of additional pieces of information: the start and end points of the load distribution. This load's beginning and ending locations occur at the two ends of the lid. As a result, you show the diagram of this load on the crate's F.B.D. as a series of downward arrows acting along the entire area of the crate's lid. Figure 13-3 shows how this distributed load is depicted on the lid of the crate.

The second part of this figure illustrates how you can also display the resultant for this load distribution. See Chapter 10 for more information about computing the resultant.

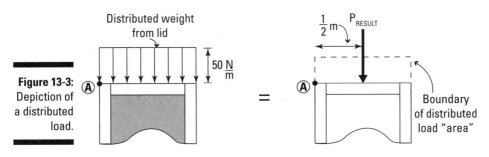

Figure 13-3: Depiction of a distributed load.

Distributed weight from lid

$50 \frac{N}{m}$

Ⓐ

=

Ⓐ

$\frac{1}{2}$ m

P_{RESULT}

Boundary of distributed load "area"

Looking at the F.B.D. so far

Figure 13-4 shows the combined F.B.D., including all the external forces I describe in the preceding sections. I've also added vertical and horizontal reactions at the contact surface to keep the free-body diagram correct. I explain more about these contact surfaces in the "Restricting Movements with Support Reactions" section a little later in this chapter.

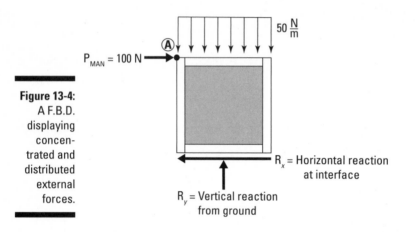

Figure 13-4:
A F.B.D.
displaying
concen-
trated and
distributed
external
forces.

Conveying concentrated moments

The third type of external load you must remember is loads created by *concentrated moments* (which cause an object to rotate). Concentrated moments are actually fairly easy to depict on a free-body diagram.

Because moments are vectors, too, they must again display magnitude, sense, and point of application. Figure 13-5a illustrates how a concentrated moment can be depicted on a free-body diagram. For both two- and three-dimensional objects, you display the magnitude of a concentrated moment as the numerical value (if it's known), or a label for the vector (if it's unknown).

The point of application and sense are usually depicted slightly differently for moments in two dimensions and moments in three dimensions, although the methods still have similarities.

Moments in two dimensions

For a two-dimensional object as shown in Figure 13-5b, the sense and point of application are determined as follows, depending on whether you're using a single- or double-headed arrow:

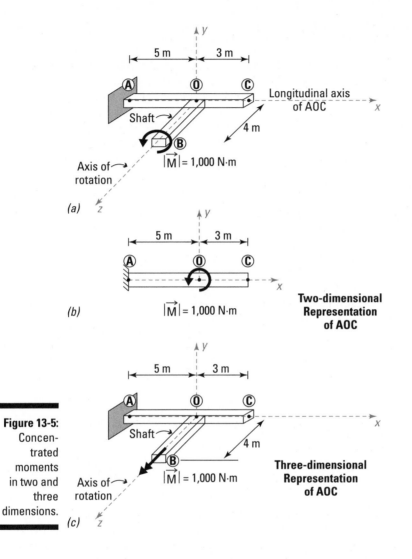

Figure 13-5: Concentrated moments in two and three dimensions.

✔ **Single-headed circular arrow:**

- **Sense:** The sense of a concentrated moment is determined by the direction of the circular rotational arrowhead. In this example, the moment is acting counterclockwise.

- **Point of application:** You can depict the concentrated moment by drawing the circular rotational arrowhead about a given point. In this case, the point of application is at Point O.

✔ **Double-headed arrow:**

- **Sense:** The sense of the double-headed arrow is determined by the right-hand rule for moments, which I explain in Chapter 12. In Figure 13-5c, the 1,000 Newton-meters (N-m) cause a counterclockwise rotation when you look at the end of the shaft. This moment can also be transmitted to Point O, as shown in Figure 13-5b, resulting in a counterclockwise applied moment on member AOC.

   Remember, you don't usually see the axis of rotation in two dimensions because it's often oriented perpendicular to the plane of the drawing (or out of the page). For two dimensions, you just use the normal circular vector depiction as I show in Chapter 12.

- **Point of application:** Just like for the concentrated loads I describe in "Portraying concentrated forces" earlier in the chapter, you can also apply the double-headed arrow with the tail or head acting at Point B. However, this point must have dimensions to help properly locate this action.

In most two-dimensional free-body diagrams, the moment acts about the z-axis (which is the Cartesian axis that seems to be coming out of the page). Double-headed arrows always act along the axis of rotation of the object, so distinguishing the sense of the double-headed arrow in two-dimensional pictures can be hard (because you can't exactly draw the arrow out from the page). For that reason, you should probably use the circular arrowheads I discuss earlier in this section to denote the direction of the applied moment (shown in Figure 13-5b). However, for cases where the moment is acting about a line in the plane of the picture (in the xy Cartesian plane), you can still use double-headed arrows.

Although you can use double-headed arrows in two dimensions, they're definitely better suited for problems in three dimensions, which I discuss in the following section.

Moments in three dimensions

For a three-dimensional free-body diagram, the sense and line of action are determined as follows:

✔ **Sense:** As with two-dimensional moments (see the preceding section), the sense of a three-dimensional concentrated moment is determined by the right-hand rule for moments (see Chapter 12).

✔ **Line of action and point of application:** For a three-dimensional rotation, you need to indicate both the line of action and the point of application on the object. The *line of action* is simply the axis of rotation about which the moment is acting, and it always passes through the point of application (Point O) on the F.B.D. as shown in Figure 13-5c. You need to be sure to include all the necessary dimensions to locate the point of application and the orientation of the axis of rotation.

Axial Loads and Beyond: Depicting Internal Forces

Although external forces (see the earlier "Displaying External Forces" section) are the easiest to define and draw, a properly drawn F.B.D. must depict all forces, including any revealed internal forces, and their locations acting on the object.

Internal forces only appear on an F.B.D. after you've sliced the object or structure — that is, when you're looking at a part of an object and not the entire object. (In Chapter 14, I explain how to know when to include an internal force.) Typically, you treat internal forces as concentrated loads and concentrated moments and draw them in the exact same manner I describe in the "Displaying External Forces" section of this chapter. Their points of application are assumed to be at the *centroid* (geometric center) of the cross section. (Chapter 11 gives you more detail on centroids.)

Restricting Movements with Support Reactions

Support reactions are the restraints that keep a rigid body from moving away when a force is applied, and they are typically classified into two categories: two-dimensional (or planar) supports and three-dimensional (or spatial) supports.

- ✔ **Two-dimensional planar supports:** *Planar support reactions* are the restraints for two-dimensional objects. Two-dimensional supports can have as many as two restraining forces and one restraining moment, depending on the type of support reaction.

- ✔ **Three-dimensional spatial supports:** *Spatial support reactions* can be much more complex. Three-dimensional restraints will have as many as six different forces and moments acting on a given support.

When dealing with support reactions, you must take into consideration the restraints to any motions on the object. If a motion is restrained, a support reaction has been created and must be included on your free-body diagram. For all support cases, you sketch the support reactions exactly as they're drawn, as concentrated forces and moments acting at the support location. Refer to "Displaying External Forces" earlier in the chapter for more on drawing concentrated forces and moments.

Restraints for *translation* (movement along a line in any direction) are always forces, and restraints for rotation are always moments.

Three basic planar support reactions

In two-dimensional statics, support restraints are categorized into one of three different support conditions: roller supports, pinned supports, and fixed supports, which I dive into in the following sections.

Rolling along with roller supports

The simplest of the three planar support reactions is the *roller support,* which is free to move parallel to the support surface and to rotate but is restrained from moving perpendicular to the support surface. Examples of roller supports include a pair of roller skates and the wheels on a car. In most textbooks, roller supports are depicted as either a single wheel, or multiple wheels as shown in Figure 13-6, where the simplest roller support is designated by a simple wheel. The object shown in this figure is free to move in one direction parallel to the support surface (or left and right) but is restrained in moving in the perpendicular. It's also free to rotate about the support location.

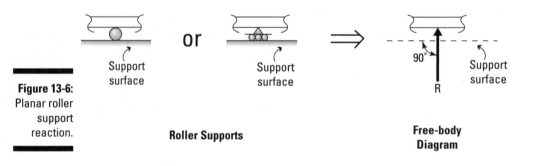

Figure 13-6:
Planar roller support reaction.

Roller Supports

Free-body Diagram

Freeing up rotation with pinned supports

The second of the three planar support reactions is known as the *pinned* or *simple support* and is a support reaction that restrains translation in two directions but is free to rotate. You commonly encounter two types of pinned supports in statics.

- ✔ **External pinned support:** As its name suggests, the *external pinned support* is a support condition that restrains an object externally. You usually depict this type of support by drawing a triangular support reaction as shown in Figure 13-7.

✔ **Internal pinned (hinge) support:** The *internal pinned support* is also known as an *internal hinge.* At this location, the internal motion of the object is restrained from translating but is free to rotate. I explain a lot more about internal pins when I talk about frames and machines in Chapter 20.

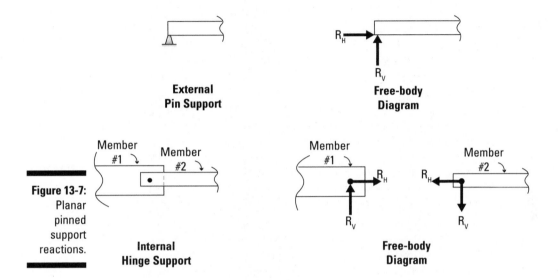

| **External Pin Support** | **Free-body Diagram** |

Figure 13-7: Planar pinned support reactions.

Internal Hinge Support **Free-body Diagram**

As with the roller supports in the preceding section, the restraint of motion is what creates the support reaction. Because this support is restrained from translating, it must have a restraining force in at least two mutually perpendicular (or *orthogonal*) directions to help hold it back or prevent it from moving. As long as these two reaction forces remain perpendicular to each other, their overall orientation doesn't matter. However, for convenience, these reactions are typically aligned with the Cartesian *x*-axis and *y*-axis of your coordinate system.

Restricting everything with fixed supports

The third type of support is known as the *fixed support reaction* in all three possible directions, as shown in Figure 13-8. Because this support restrains both of the translations in addition to the rotation, your drawing must have three separate support reactions.

Figure 13-8: Planar fixed support reaction.

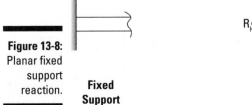

Fixed Support

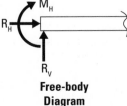

Free-body Diagram

As I mention with the pinned supports in the preceding section, the two translation restraints produce two restraining forces. However, for fixed supports, the rotational behavior of the support is also restrained, and thus you must also include a concentrated moment restraint.

Moving on up (or down) with inclined supports

Inclined supports can be either roller, pinned, or fixed supports; the supports are just no longer horizontal or vertical but rather inclined along a slope. Keep these rules in mind when you're dealing with inclined supports in two dimensions:

- ✔ If a support is free to move, no restraining force is developed.

- ✔ If the object is free to rotate, no restraining moment is developed.

- ✔ If the object is restrained from rotation, the restraining moment that's created is always about an axis perpendicular to the plane of the object.

Roller supports don't always have to be aligned horizontally or vertically. In fact, rollers can be oriented in any direction, so you want to pay careful attention to which way the supporting surface is oriented with respect to the free-body diagram. Figure 13-9 shows a roller support oriented on a surface inclined at an angle θ measured from the horizontal.

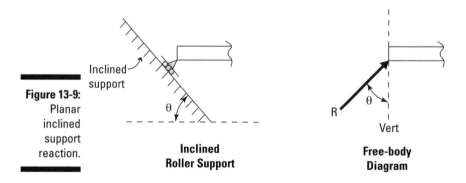

Figure 13-9: Planar inclined support reaction.

Inclined Roller Support

Free-body Diagram

In this example, the roller support is free to move down the incline, so no force is acting parallel to the incline. However, the restraining force R of the supporting surface prevents the motion perpendicular to the surface. With these two directions accounted for, the behavior in the *xy* plane is now properly defined.

Because the supporting surface prevents the object from moving in that direction, a hidden support force (or a support reaction) must be created to prevent the object from moving. In this case, the support reaction is oriented perpendicular (or *normal*) to the support surface.

Three-dimensional support conditions

Three-dimensional supports can be a bit more complicated than their two-dimensional counterparts (see the preceding section). In two-dimensional analysis, three support reactions develop at most: two translational forces (*x*- and *y*-directions) and one rotational moment (*z*-direction).

To fully define translation in three dimensions, you must define three mutually perpendicular forces, typically with respect to the Cartesian *x*-, *y*-, and *z*-axes. Similarly, you must also account for three mutually perpendicular moments to define rotational restraint in three dimensions.

To model three-dimensional support reactions, you have to once again apply a bit of logic about how the object is capable of moving. If any restraint is present (either translation or rotation), you must include a support reaction to represent that restraint. Although countless three-dimensional supports are possible, I describe a couple of the more common ones in the following sections.

Ye olde ball and socket: Pinned supports in three dimensions

A three-dimensional pinned support is commonly referred to as a *ball-and-socket connection* and is shown in Figure 13-10. (See "Freeing up rotation with pinned supports" earlier in the chapter for pinned-support basics.) Examples of common ball-and-socket supports are the hip and shoulder joints on your body. If you try moving your arm around, you notice that with some effort, you should be able to rotate your arm in any of three directions, but your shoulder stays in the same place. That is, it's free to rotate in any direction but is fully restrained from translation (or popping out of its socket). Because restraint causes support reactions, you know that your shoulder has three translational support reactions but zero rotational support reactions.

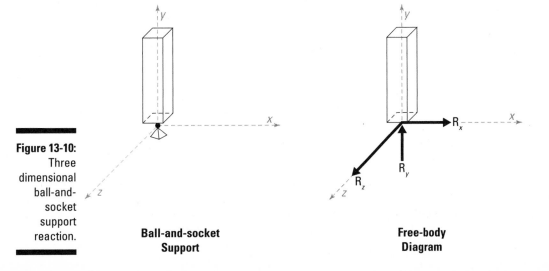

Figure 13-10:
Three dimensional ball-and-socket support reaction.

Ball-and-socket Support

Free-body Diagram

Collar assembly supports

Another common support reaction that occurs in mechanics is the *slider* or *collar assembly.* This type of support consists of a sleeve that wraps around a rod or shaft. In this connection, the sleeve is free to translate parallel to the axis of the rod, and is capable of swiveling (or rotating) about the axis of the shaft. All other motion (both rotational and translational) is restrained. Figure 13-11 shows a common collar assembly and its corresponding free-body diagram.

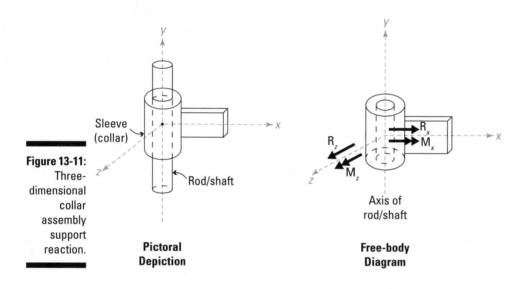

Figure 13-11: Three-dimensional collar assembly support reaction.

Pictoral Depiction

Free-body Diagram

As drawn in this example, the shaft is oriented along the y-axis. Because this assembly is free to move in the direction of the shaft, it's considered unrestrained in that direction and should have no force for the support reaction in that direction. Similarly, the sleeve is capable of rotating about the axis of the shaft (the y-axis) and consequently has no moment for the support reaction about the y-axis. The remaining motions are all fully restrained and consequently have both moments and forces as support reactions in the x- and z-directions.

Weighing In with Self Weight

The fourth force category deals with the representation of the self weight of the object. In Chapter 9, I explain how to calculate self weight for a lumped mass (concentrated system), which is a concentrated force acting at the object's center of mass. However, if you're dealing with a distributed mass system (as I describe in Chapter 10), the self weight is actually a distributed force. In statics, you commonly neglect the self weight of an object unless the problem explicitly states that the object has mass. This omission is usually acceptable because the nature of many structural systems is such that the force from self weight is usually only a small percentage of the total force acting on a system. For example, a small construction crane that weighs only a few thousand pounds is often capable of lifting loads of many hundreds of thousands of pounds. That being said, you absolutely want to consider the exact self weight of the object in addition to the external loads if you're performing a final design of the crane.

For simplicity in this book, I often neglect self weight. If I don't mention the mass of the object, I also exclude the self weight from the F.B.D. and subsequent calculations.

Chapter 14

The F.B.D.: Knowing What to Draw and How to Draw It

A *free-body diagram* (F.B.D.) is the physical representation of an object (or part of an object) with all actions acting on it. In statics, this setup usually includes forces such as concentrated and distributed loads, support reactions, internal forces, and self weight; check out Chapter 13 for more on these force categories. These four categories of forces help depict all forces on a given object (although in non-statics problems, you may include other vector actions such as velocities, accelerations, distances, and many others). Consider an 8,000-pound elephant and an 8,000-pound commercial vehicle. Both have the same *self weight* (force due to gravitational effects on their masses), but the location of that weight tremendously affects the objects they're resting on in drastically different ways.

In this chapter, I show you some of the situations you need to be mindful of when tackling any statics analysis problem with an F.B.D. I start with a basic list of items you want to be sure to include on your free-body diagram. I also show you how you can extract additional free-body diagrams from within a given F.B.D.

Getting Your F.B.D. Started

The old saying that a picture is worth a thousand words may seem clichéd at first, but this simple statement accurately emphasizes the importance of a well-constructed graphical representation for statics analysis. In fact, without this pictorial description, you quickly discover that accurate engineering and physics solutions are next to impossible to achieve. In fact, even a simple photograph can sometimes serve as a basis for constructing a free-body diagram.

Think of an F.B.D. as being a snapshot of the support reactions and forces acting on an object at a particular instance.

To construct an F.B.D. in Cartesian components by starting with a basic picture, just follow these steps:

1. **Sketch the complete object that you're going to be studying, such as a chair or a ladder.**

   In reality, the actual shape of the object really doesn't matter. However, knowing the exact locations *(points of application)* of every force and moment vector on that object is crucial. Actually drawing the object can give you a reference for measuring those locations.

2. **Draw all external *support reactions* (physical movement restraints) as well as any springs or cables that are attached to the system.**

   Support reactions include pinned supports, ball-and-sockets, slider assemblies, and so on; I discuss them in more detail in Chapter 13, where I also show you how to draw basic two- and three-dimensional support reactions. Springs and cables are generally fairly easy to identify and can provide a quick confidence boost in the construction process; see Chapter 9 for more on these items.

3. **Include complex supports, such as ramps and inclined supports.**

   Examples of these additional restraints include refrigerators on ramps and wedges under crates, along with any information about inclined support surfaces (usually expressed in degrees). Flip to Chapter 13 for more on inclined supports. In fact, if your object is resting on a ramp or incline, be sure to include the angle of the incline as a reference, even if you're focused on just the object.

4. **Draw each external force acting on the systems, as described in Chapter 13.**

   Assume that all objects in the F.B.D. are *rigid* (meaning they aren't deformed by the force), even if the system has springs; springs aren't actually rigid, but you can consider a spring rigid at the exact instance that you draw the free-body diagram.

   If your F.B.D. is of an isolated object that you're separating from a larger system or support, you also need to include *contact forces,* which are forces that arise when one object pushes on either another object or a support surface. However, you don't include contact forces on a free-body diagram when both objects (or supports) on either side of the contact surface are already included in the diagram.

Assuming a direction for support reactions

When you're drawing an F.B.D., getting the forces onto the diagram is more important than worrying about their sign or the direction. As you discover in Chapter 16, until you start writing equations, you may not actually know the magnitude and sense of the vector for the unknown magnitudes at the time you're drawing the free-body diagram. If you don't know the *magnitude* (length) and *sense* (direction) of a vector when you draw it, simply guess a direction, include a label on the force, and apply it at its proper point of application along its appropriate line of action.

For example, say you want to apply the support reactions to a diagram of Figure 14-1a, which illustrates a man standing at the end of a diving board. Based on the sketch, you may be inclined to draw the support reactions as positive with respect to your Cartesian coordinate system as shown in Figure 14-1b, with all vertical loads acting upward and all horizontal loads acting to the left. And while being consistent is good practice for now, you may soon realize that it's not always correct. In fact, Figure 14-1c actually represents the correct F.B.D. of Figure 14-1a. Notice that the difference between Figure 14-1b and c is only in the direction of the vertical reaction A_y located at Support A — upward in b but downward in c.

You always know the direction of a load resulting from self weight – it's acting downward and applied at the center of mass, so be sure to go ahead and draw it in the correct direction.

Including more than the required info on your F.B.D.

In Part II, I emphasize the importance of being able to quickly create Cartesian vectors, which comes in handy when you realize that you often repeat the vector creation process many times on the same free-body diagram. To make your work easier later, get in the habit of including several additional pieces of information, including the following:

> ✔ **Coordinate axes:** The *coordinate axes* are a useful reminder of any assumptions you've made about positive and negative directions. Including information regarding the positive directions of each of the *x*-, *y*-, and *z*-directions is especially useful because when you start writing the equations of equilibrium (which I introduce in Chapter 16), these directions can help you determine the sense of any unknown vectors.

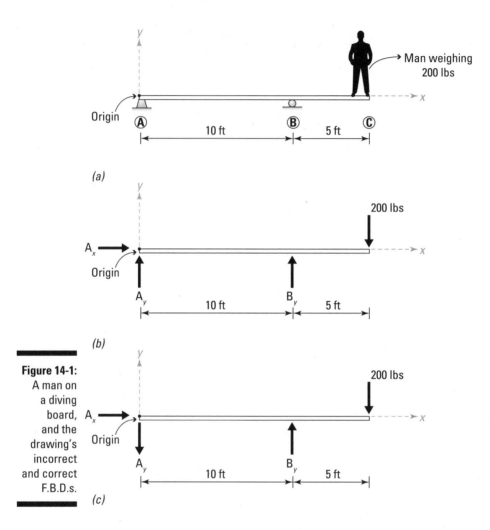

Figure 14-1:
A man on
a diving
board,
and the
drawing's
incorrect
and correct
F.B.D.s.

Although you typically take the Cartesian *x*-axis as being horizontal (and consequently the Cartesian *y*-axis as vertical), sometimes aligning your axis in some other direction is more convenient. You encounter this situation often when working with problems on ramps and inclined surfaces, or in problems that involve forces acting at some non-Cartesian orientation (or at some angle other than horizontal or vertical).

✔ **Origin:** The *origin* is the location where the Cartesian axes for your system intersect; it serves as a handy reference location and customarily has the coordinates (0,0,0).

✔ **Labels for points of interest:** You also should label some key points of interest on your diagram such as the following:

- All support and internal hinge locations (see Chapter 13)

- All locations of applied external concentrated loads (see Chapter 13)

- All locations where forces' lines of action cross the object or other lines of action (see Chapter 4)

- All resultant force points of application (see Chapter 7)

- All locations at the start and end of the free-body diagram (see Chapter 10)

- All changes in the geometry of the object

- All centers of gravity and centers of mass (see Chapter 11)

✔ **Force vector components:** You can also use dotted arrows to indicate perpendicular components and their senses with respect to your assumed Cartesian axis direction. Calculate the components of the force vectors and sketch them as individual forces on the free-body diagram.

✔ **Dimensions and angles of supports and forces:** You also want to include dimensions that relate all of the locations discussed in the preceding bullets. Be sure to also include information such as angles or proportion triangles that may help describe the orientation of any lines of action for forces on your free-body diagram. (Chapter 5 gives you the skinny on proportion triangles.) You'll also need to include the angles of any inclined supports (see Chapter 13).

If you provide adequate linear dimensions, angular dimensions may not be necessary. A little trigonometry can be used to compute angles on free-body diagrams.

Zooming In with Isolation Boxes

Isolation boxes let you take larger objects and zoom in on specific features. An isolation box tells you when an internal force needs to be included on a free-body diagram — whenever an isolation box crosses a physical object, you must include an internal force on the object.

Several features that you may want to explore in further detail on a given free-body diagram include support reactions and internal forces in members. The following sections show you how you can use an isolation box to extract smaller (and sometimes more manageable) pieces of a free-body diagram while preserving the overall behavior of both the larger system and the isolated portion.

Unveiling internal forces

Depending how you cut the picture when you draw a free-body diagram, you can greatly reduce or increase the complexity of the system. For example, Figure 14-2a shows a system of two cables oriented at different directions connected to a ring that's suspending a 100-kilogram box.

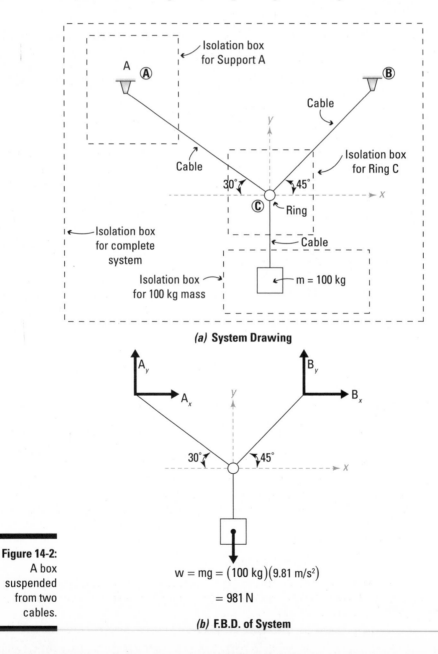

Figure 14-2: A box suspended from two cables.

(a) **System Drawing**

(b) **F.B.D. of System**

According to the procedure outlined in "Getting Your F.B.D. Started" earlier in this chapter, the first diagram you should draw is the free-body diagram of the entire system, which I show in Figure 14-2b. The system has two support reactions each at location A and location B generated by the pinned supports. The only load applied to the system is the self weight of the box, which has a force of 981 Newton (100 kilograms · 9.81 meters per second squared = 981 Newton).

A total of five forces are acting on the system in Figure 14-2a, four of which are unknown at this time. You won't actually calculate the unknown forces at this time (I explain how to do that beginning in Chapter 16), but I do show you the free-body diagrams you need to create as part of this process.

In this system, suppose you're interested in calculating the internal forces in each of the rope sections. In order to see these internal forces, you need to create isolation boxes that cut these objects and expose these forces. Because all the connecting objects are cables, the internal forces are rather simple and include only axial forces. The sense of the internal cable forces is determined from the knowledge that cables are only capable of transmitting tension (or pulling on the free-body diagram). I discuss more about cable requirements in Chapter 22.

Applying rules of application

The main free-body diagram of interest in Figure 14-2a is at the ring at location C, where all three cables meet. To draw the free-body diagram of just the ring at C, you use an isolation box around the ring.

To construct an isolation box, stick to the following basic steps:

1. **Identify the object or detail of interest.**

   Decide what part of a structure you need to examine in further detail. In this example, you're interested in what's happening at the ring at location C.

2. **Construct a closed polygon around the object or detail of interest.**

   Draw a box, circle, or some closed polygon shape — usually with dashed lines — to differentiate between the isolation box and the physical object itself (refer to Figure 14-2a). Make sure that the isolation box is a closed polygon. Labeling the box is handy if you're going to be using multiple isolation boxes.

3. **Extract the contents of the isolation box.**

   Copy everything inside the isolation box to a new picture. Include external forces, supports, self weight, and all relevant dimensions and Cartesian coordinate system data.

4. **Include all internal forces revealed as a result of the cutting process.**

The chapters of Part VI deal with structures with different types of internal forces. Depending on the type of system you are working with, different internal forces may appear when you cut an object. For this example, you're dealing with a system of axially loaded cables, which I introduce in Chapter 9.

Look at the free-body diagram again. At any location where the isolation box crosses a physical object, internal forces are revealed and must be included. In this example, the isolation box for the ring at location C cuts three different cables, so you must apply a concentrated axial force at each of these locations.

The isolation box for the ring at location C cuts each of the three cables, exposing their internal forces on the free-body diagram. Figure 14-3 illustrates what a properly constructed free-body diagram for the ring at C looks like. Notice how even though you don't know the magnitude of the internal force, you do know the sense of the force because cable forces always act in line with the cable and they're always in tension (or pulling on the object).

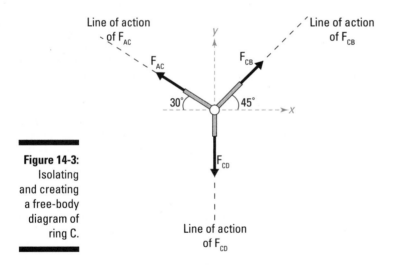

Figure 14-3:
Isolating and creating a free-body diagram of ring C.

Another object of interest in this example may be the box itself. Because you already know the mass of the box, you know that there's a self weight force present on this system. Sketching an isolation polygon around the box requires that you also cut cable CD. As before, this move results in an unknown internal cable force pulling on the box at D as shown in Figure 14-4.

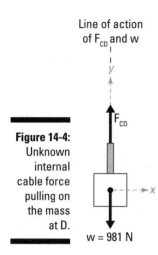

Figure 14-4:
Unknown internal cable force pulling on the mass at D.

Take a look at the force vector $\mathbf{F_{CD}}$ in Figures 14-3 and 14-4. Notice that the internal force in the cable CD is oriented in one direction in 14-3 and the opposite in 14-4. Both represent tension in the cable because they're pulling on their respective free-body diagrams. However, the sense of the force changes depending on which isolation box you're working with. In this case, that's perfectly acceptable. Just make sure that internal forces on opposite sides of a *cut line* (or the location where an isolation box crosses the physical object) are equal and opposite in magnitude, sense, and direction.

You can also use an isolation box to capture unknown support reactions and relate them to internal forces. On this diagram, you include the two unknown support reactions A_x and A_y and the unknown revealed cable force F_{AC} with a 30-degree dimension indicating the orientation of the cable (and as a result, its internal force) as shown in Figure 14-5.

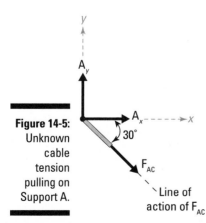

Figure 14-5:
Unknown cable tension pulling on Support A.

Avoiding problems with incorrect isolation techniques

The art of using isolation boxes takes some trial and error to get used to, but after you get the basics down, they provide the framework for quickly moving through a system with calculation and equation techniques that I discuss in later chapters.

One of the potential problems of isolation boxes is the increase in the number of unknown reactions or internal forces that you must include on the free-body diagram. For example, consider the simple truss structure shown in Figure 14-6a, which already indicates the support reactions. If you use an isolation box to cut only members CD, CG, and HD as shown in Figure 14-6b, notice how the complexity of the free-body diagram greatly increases. You've cut each member at two different places, exposing six additional internal forces as shown in Figure 14-6b — one on each side of each member within the isolation box.

The free-body diagram of the remaining portion of the structure also has these same six revealed internal forces (as shown in Figure 14-6c). However, each pair of forces is acting along the same line of action, with one of the forces acting in one direction and the other acting in the opposite direction for each member. Vector addition tells you that the net effect on that member is zero (for example, $F_{CD} - F_{CD} = 0$).

Your isolation box hasn't really yielded a whole lot of new information. What you want to do is cut the objects in your free-body diagram in a way such that only one of any unknown internal force is present on any free-body diagram. The simplest way to accomplish this feat is to require cutting all the way through the object or system. Figure 14-7 illustrates a better way of cutting this structure, resulting in only one of each unknown internal force on the system.

In this example, notice that you only have four unknown forces to deal with, whereas before you had six — all of which cancelled each other because they were acting in opposite directions with the same magnitude along the same line of action.

By cutting the system entirely into two pieces with your isolation box, you actually produce two free-body diagrams with the same unknown internal member forces. This result is useful because when you start writing equilibrium equations (which I show in Chapter 16), you can use either of the free-body diagrams. Sometimes, one free-body diagram is significantly simpler, meaning it has fewer loads, fewer reactions, or more-convenient geometry.

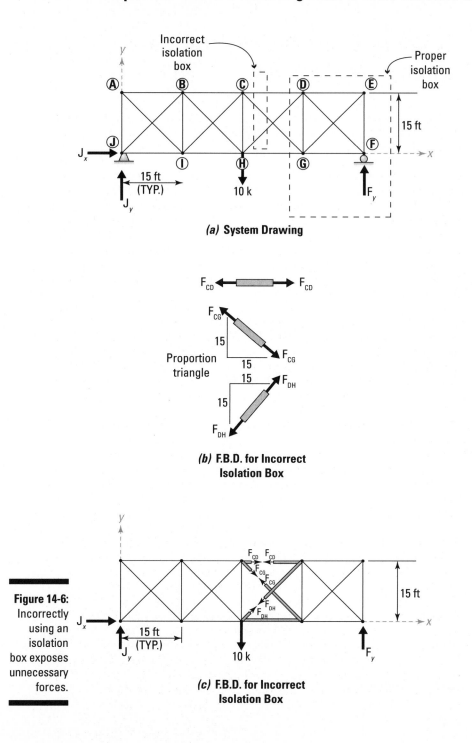

(a) **System Drawing**

(b) **F.B.D. for Incorrect Isolation Box**

Figure 14-6: Incorrectly using an isolation box exposes unnecessary forces.

(c) **F.B.D. for Incorrect Isolation Box**

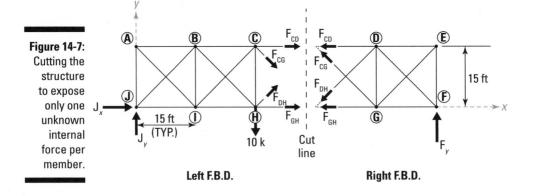

Figure 14-7:
Cutting the structure to expose only one unknown internal force per member.

Left F.B.D.   Right F.B.D.

Using Multiple F.B.D.s

When multiple objects interact with each other, forces inherently exist between the objects. The chair you're sitting on applies a force to the carpeted floor below (if you don't have carpet, pretend you do); you can see the effect when you move the chair and see the impression of the chair legs that remains behind. Likewise, the carpeted floor itself exerts a force on the legs of the chair, preventing it from falling through the floor into your basement.

In systems with multiple objects connected, such as with a clamp or pair of pliers, looking at each of the individual pieces is useful. When *exploding* an object (separating connected pieces), your isolation box typically has to cut through a pin, bolt, or some other sort of connecting element. Cutting through pins or connections is one of the tricks I explore more in detail in Chapter 21.

In problems where one object is resting on top of another, such as the block resting on a ramp as shown in Figure 14-8a, the forces interacting between the two objects are what tell the story.

The only forces acting on the system as drawn are the weight of the block itself and the external applied load P, which is 200 Newton. So if the block is sitting on an incline and an applied load is pushing the block down the incline, what's holding the block in place? The answer is *friction,* which is an invisible force that exists between two objects as they attempt to move past each other. (Don't worry; I discuss more about friction in Chapter 24).

When you draw two isolation boxes, the internal balancing forces actually show up. Figure 14-8b shows the free-body diagram of just the block. In this picture, if you draw only the 200-Newton applied force and the 150-Newton self weight, you see that the block would clearly want to move downward and to the left as drawn.

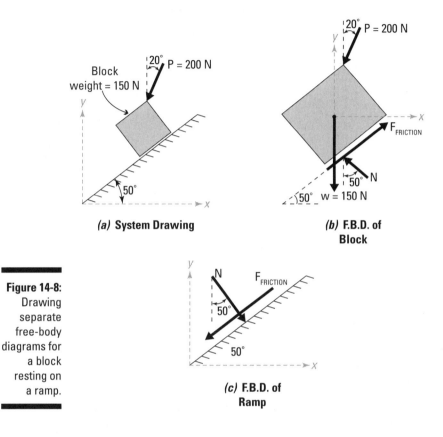

(a) **System Drawing**

(b) **F.B.D. of Block**

(c) **F.B.D. of Ramp**

Figure 14-8: Drawing separate free-body diagrams for a block resting on a ramp.

However, this scenario can't physically happen. Clearly, the ramp has to be pushing back on the block; otherwise, the block would fall through the ramp, which is physically impossible. Think of the support reaction of the block as a quasi-roller support (sort of like the roller supports I discuss in Chapter 13). To counter the downward forces, a vertical supporting force also has to be present to maintain the block in its original position. A portion of this resistance is due to the normal force N.

To prevent the block from sliding to the left, a second force has to be introduced that has a component pushing to the right. The only way for this situation to occur is for a force vector to be developed that's acting up the ramp. To balance these two forces that are acting on the block at the interface of the block and ramp, a second set of forces equal in magnitude but opposite in sense have to be acting on the ramp as shown in Figure 14-8c.

Chapter 15

Simplifying a Free-Body Diagram

. . . .

In This Chapter

▶ Understanding the principle of superposition

▶ Relocating forces and moments

▶ Determining equivalent systems with the space potato analogy

. . . .

Chapters 13 and 14 equip you with the countless tools and equations to create, calculate, and depict the behaviors of vectors on objects by using free-body diagrams (F.B.Ds). F.B.D.s can be complex, intimidating, and a bit overwhelming because you often have an object that is subjected to distributed loads, concentrated forces, and applied moments applied at multiple locations, all acting simultaneously. (Flip to the chapters of Part III for more on these concepts.) Figuring out where to begin in tackling these diagrams can be quite a task.

One of the first steps you want to perform after making your sketch is to look for ways to simplify your F.B.D. In this chapter, I show you several handy simplifying techniques that you can utilize on a regular basis.

Presenting the Principle of Superposition

The *principle of superposition* basically states that multiple actions on an object are equivalent to the sum of the effects of each action applied individually. The principle of superposition allows you to quickly compute behaviors (such as reactions, displacements, and internal forces) from combined multiple load cases by simply adding together the responses of the individual cases.

I assume that all objects in this book are *rigid bodies,* which are objects that aren't deformed by the forces acting against them. Although the principle of superposition is based on assumptions surrounding *non-rigid bodies* (bodies transformed by forces), it's still a handy tool in mechanics and statics because it provides good approximations for many rigid bodies.

Making the assumptions for superposition

To properly utilize the principle of superposition, I first need to explain several fundamental assumptions that actually appear frequently in the derivations and theorems of mechanics and physics: small displacement theory, linear system behavior, and elastic material behavior.

The basic assumptions that you need to be mindful of are when a load acts on an object, the displacement behavior of the object due to the load must remain very, very small. In most classical structures, this isn't an issue because most real-world structures experience very small displacements. For the purposes of this book, you can assume that all of these requirements are met.

The remaining assumptions are mostly related to deformation and deformable bodies. However, in this text, you are dealing exclusively with rigid bodies. If you want more details about these common assumptions, check out any basic mechanics of materials or strength of materials textbook.

The major assumptions behind static analysis naturally satisfy the requirements for applying the principle of superposition. Because you're dealing with rigid bodies, a common assumption is that small displacements are actually zero displacements. Assuming zero displacement in a system also automatically satisfies the elastic material and linear system behavior assumptions because they're based on zero displacement.

Most static systems in this text are linear systems, with the exception of cable systems, which are almost always nonlinear. Don't worry! I explain more about cable systems and how to deal with them in Chapter 22. For all examples in this book, I assume that small displacement theory is valid and that the materials of the objects in each example behave elastically (or return to their original shapes when the load is removed).

The fundamental principles behind the principle of superposition have their origins in vector formulations for finding resultants. In Chapter 7, I show you how to take two vector components, $\mathbf{F_1}$ and $\mathbf{F_2}$, and simply compute the vector sum to find a resultant vector $\mathbf{F_R}$. If you have more than one vector component, you can add as many as you like to find the resultant vector. That is:

$$\vec{F_R} = \vec{F_1} + \vec{F_2} + \dots$$

The idea of determining each part individually and then combining them to find a single combined behavior is the premise behind the principle of superposition.

If you consider all the assumptions from the preceding sections to be true for Figure 15-1, you can apply the principle of superposition by examining the effects of each load individually, creating three separate F.B.D.s that reflect the behavior of each of the individual loads.

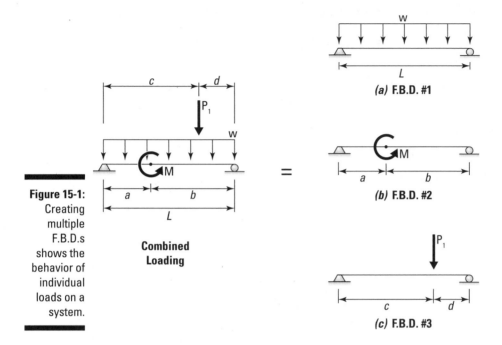

Figure 15-1:
Creating
multiple
F.B.D.s
shows the
behavior of
individual
loads on a
system.

(a) F.B.D. #1

(b) F.B.D. #2

(c) F.B.D. #3

**Combined
Loading**

The combined loading of this example consists of a simply supported beam subjected to a concentrated load $\mathbf{P_1}$ acting at a distance c from the left support. In the same loading, a uniform distributed load with a magnitude of w is applied for the entire length of the beam. Finally, a concentrated moment is applied at a distance of a from the left end of the beam.

For this example, F.B.D. #1 (shown in Figure 15-1a) is a simply supported beam subjected to a uniformly distributed load $\mathbf{w}$ over the entire length. F.B.D. #2 (Figure 15-1b) is the same beam and support reactions with a concentrated moment $\mathbf{M}$ at a distance a from the left support. F.B.D. #3 (Figure 15-1c) is the same beam and support reactions yet again, with the concentrated load $\mathbf{P_1}$ applied at a distance c from the left support. The combined effect of each of these three load diagrams results in the same combined loading.

This principle is extremely useful for complex loadings situations in statics because it allows you to break a problem down into more manageable pieces. In fact, you may be surprised to find that many design handbooks out there have simple load cases already worked out.

Centering on Centerlines and Lines of Symmetry

Another useful simplification technique is to look for a line of symmetry within a structure or loading diagram. A *line of symmetry* is an imaginary line that produces a mirror image of one part of a load or structure onto another. For a symmetric condition to occur, you must make sure that each of the following is mirrored by the line of symmetry:

- **Beam properties:** Properties such as mass, geometry, and cross-sectional properties.

- **Support reactions:** *Support reactions* (restraints that keep a rigid body from moving away when force is applied; check out Chapter 13 for more). A fixed support condition on one side of the mirror line must be reflected on the other. The one exception to this rule is with pinned and roller supports. If all loads are in one direction, a roller support can actually mirror a pinned support as long as the parallel component of the pinned support is zero.

- **Load magnitude and location:** A distributed load must be mirrored by either being centered on the line of symmetry or by having an identical distributed load on the reflected side of the line. Similarly, you can consider a concentrated load or moment symmetric by itself if it occurs on the line of symmetry; otherwise, it must have a matching concentrated load or moment and location that is reflected in the line of symmetry.

If any one of these items isn't properly mirrored about the line of symmetry, you can't consider the loading symmetrical, and the methods in this section won't always work.

In Figure 15-2a, you see a simply supported beam with a partial uniform distributed load of magnitude w, with a distributed length of $2b$. Because the load is centered on the beam, it's symmetrically loaded — if you draw a mirror image line (or line of symmetry) exactly in the middle of the beam, the loads, supports, and beam conditions on the left half will be exactly the same as those on the right. In cases involving symmetry, you can usually assume that the reaction's symmetric loading will be one half of the total resultant of the symmetric loading. In this case, the total applied load amounts to a resultant of $2wb$, or wb for each reaction $\mathbf{R_A}$ and $\mathbf{R_B}$.

Conversely, Figure 15-2b shows the same load applied nonsymmetrically. In this example, no line of symmetry meets all the requirements for mirroring of supports, loads, and geometry.

In some cases, lines of symmetry can also occur at different angles from the rest of the structure, as shown in the loading of the L-shaped object of Figure 15-2c. In this example, the loads and geometry aren't mirrored by a vertical or horizontal line of symmetry but rather on a line oriented at 45 degrees passing through the corner of the L shape.

TIP

Symmetry allows you to use the principle of superposition to quickly determine information about an F.B.D. without having to write any of the more complex equilibrium equations I show you in Part V. For example, consider the beam of Figure 15-2a again. With a bit of simple logic and Chapter 10's resultant techniques for distributed loads, you can quickly determine the unknown support reactions R_A and R_B. In this example, you have a partially uniformly distributed load acting in the middle of a beam. If a beam is symmetric, and the loads acting on it are also symmetrically positioned, you know right away that each support of the beam carries exactly half of the symmetric load. In this example, the total load (or resultant) of the distribution is equal to $R = 2wb$. If the force is shared equally by the supports because of symmetry, you know that the support reactions are

$$R_A = R_B = \frac{1}{2}(2wb) = wb$$

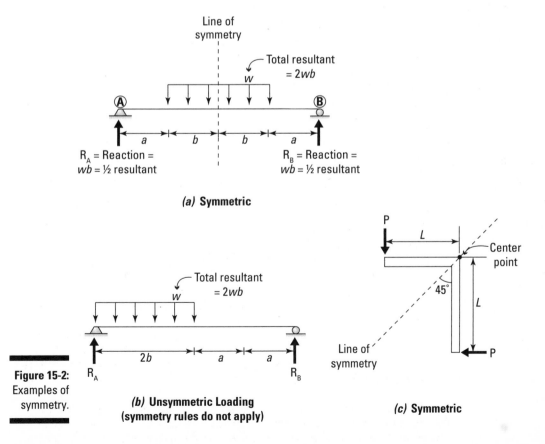

Line of symmetry

Total resultant
= $2wb$

w

(A)   (B)

a b b a

R_A = Reaction =
wb = ½ resultant

R_B = Reaction =
wb = ½ resultant

(a) Symmetric

Total resultant
= $2wb$

w

$2b$ a a

R_A R_B

**(b) Unsymmetric Loading
(symmetry rules do not apply)**

P

L

Center point

45°

L

Line of
symmetry

P

(c) Symmetric

Figure 15-2:
Examples of
symmetry.

Equivalent Systems: Forces on the Move

Another method of simplifying a structure is by determining equivalent behaviors through the relocation of forces and moments. (In fact, this principle is what drives a lot of the discussion on equilibrium beginning in Chapter 16.) An *equivalent system* is a system of forces and/or moments that you can replace with a different set of forces and/or moments and still achieve the same basic translational and rotational behaviors. (For more on these behavioral concepts, take a look at Chapter 12.)

To start the discussion of relocating a force, I now introduce you to the space potato.

Moving a force: The space potato analogy

The *space potato* is a unique, albeit somewhat nonsensical, method of referring to an arbitrary three-dimensional object in space. Suppose you have your very own space potato with a force F_A acting at a Point A somewhere on the object, as shown in Figure 15-3a.

Now suppose you want to compute the effects of this force at a different location, such as Point B, on the space potato. To determine these effects, you need to compute the equivalent system at that point.

To determine the required translational effect, add a new negative force $-F_A$ at Point A, and an additional force F_A at the force's new location (B). These two forces are illustrated by the dashed arrows in Figure 15-3b. The resultant of these newly added forces is $F_A + (-F_A) = 0$, or zero net translational effect.

The original force F_A plus the new negative force $-F_A$ also cancel each other at the original Point A. That is, the forces acting at Point A result in a net translational effect of zero at Point A. With that in mind, you can see that the only force that now remains on the space potato is the newly relocated force F_A, which is now acting at Point B. Congratulations! You've relocated your first force. Unfortunately, you've also managed to introduce a new behavior to the potato in the process.

Now that you've relocated the force, examine the two additional forces that were added in Figure 15-3b. For now, check out each of the two additional forces and their points of application as shown in Figure 15-3c. In this sketch, you see that these two new forces are a negative force $-F_A$ acting at Point A, and a positive force F_A acting at Point B. The two forces are parallel, acting in opposite directions and separated by a distance. This setup is the very definition of a moment couple that I discuss in Chapter 12. Remember, moments and couples cause rotation in an object, and the added rotational effect of this couple is what you also need to include when you move a force.

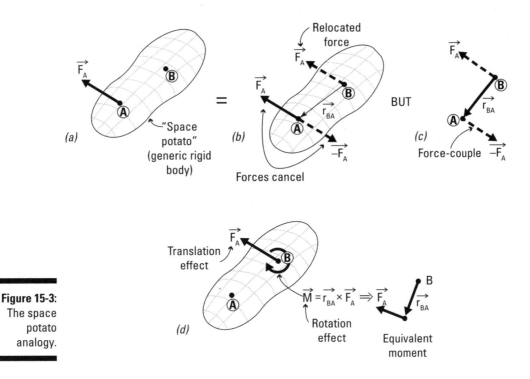

Figure 15-3:
The space
potato
analogy.

To compensate for this couple, you need to compute an equivalent moment at Point B, which requires a position vector from your point of interest to a point on the line of action (usually taken as the point of application) of the original force. Compute the equivalent moment at Point B from the basic moment vector formula:

$$\overrightarrow{M_B} = \overrightarrow{r_{BA}} \times \overrightarrow{F_A}$$

The space potato analogy shows you that to relocate a force, you simply need to take the original force and apply it at the new location, plus compute a newly applied moment and apply that at the new location. The final result of the relocated force and newly created moment are shown in Figure 15-3d.

Equivalent systems must have both the same translational behavior and the same rotational behavior at a given point. Forces provide the translational behavior, and moments provide the rotational behavior.

To relocate a distributed force, you simply need to convert the distributed force to a single concentrated resultant, determine the location of that resultant on your space potato, and then follow these same rules for relocating a force.

Moving a moment

In the preceding section, I illustrate how you can convert a couple into a single concentrated moment at a new location. But how do you handle an applied concentrated moment? As I discuss in Chapter 12, a concentrated moment is an action that causes a rotational behavior but doesn't affect the translation of an object. *Moment vectors* are a type of vector known as a *free vector,* which applies the same rotational behavior regardless of where on the object it's acting (see Chapter 4). As a result, you can freely move any moment (both concentrated and couples) around the object as long as the magnitude and sense of the moment vector remain unchanged (see Figure 15-4). The point of application of a moment or couple doesn't matter when creating an equivalent system. If you want to move a moment, just move it!

Figure 15-4:
Relocation
of con-
centrated
moments.

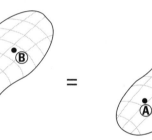

Part V

A Question of Balance: Equilibrium

The 5th Wave
By Rich Tennant

"This is my old statics teacher, Mr. Wendt, his wife, Doris, and their two children, Cartesia and Torque."

In this part . . .

This part is all about *equilibrium,* the state of balance between opposing forces. Here, I discuss Newton's basic laws of motion and their applications in statics. I provide the basic equilibrium formulas and show you how to compute unknown support reactions from free-body diagrams. Plus, I also explain the differences between two- and three-dimensional techniques for evaluating equilibrium conditions.

Chapter 16

Mr. Newton Has Entered the Building: The Basics of Equilibrium

In This Chapter

▶ Defining translation and rotation

▶ Establishing equilibrium equations

▶ Describing improper constraints

*Y*ou're sitting comfortably at your desk, intently reading about all the gooey goodness that is statics, discovering how to draw and calculate vectors, and developing free-body diagrams to describe the world around you, when you lean back to stretch your spine. In the process, you suddenly find yourself losing balance — both you and your chair fall to the ground. Before you know it, you're lying on the ground wondering, "What the heck just happened?" Your personal weight and the weight of the chair have not changed since you sat down. You, my friend, have fallen victim to the perils of equilibrium.

After you have the basics of vectors and F.B.D.s down, you're ready to start exploring the single most important concept in statics: equilibrium (or balance). In this chapter, I show you how equilibrium is defined and introduce you to some of the work of Sir Isaac Newton, who provides you with the necessary tools used to explain equilibrium.

For even more information on equilibrium, turn to Chapters 17 and 18. Chapter 17 covers scalar methods, and Chapter 18 covers vector methods.

Defining Equilibrium for Statics

The word *equilibrium* has several different meanings, but in statics lingo it's basically defined as "a state of rest." In particular, it means that an object or system isn't experiencing any motion or acceleration. Now, all you need to know is how motion is defined.

In statics, *motion* is classified into two major categories:

- ✔ **Translation:** A linear or straight-line movement of an object
- ✔ **Rotation:** A spinning or turning of an object about a reference point or axis

Translational equilibrium

Any behavior that changes an object's relative position without causing it to rotate is a called a *translational* effect. Translation is also sometimes referred to as *displacement.*

Translation occurs only when an object is subjected to unbalanced forces. That is, a net unbalanced force (or the resultant of a system of forces) must exist in at least one direction for an object to translate in that direction. In order for an object to be in a state of translational equilibrium, the resultant of all forces on the system must be zero — that is, all the forces on an object must be balanced.

Consider the object of Figure 16-1, which is experiencing translation. To properly depict an object experiencing pure translation, you need to choose two arbitrary points — say, the reference point inside the object, and another corner point, such as Point D, on the boundary. For the first point you select, trace a line from the original position of that point to its final position. In this example, I label this line "Line of translation #1." Now repeat this process for a second point (such as Point D). For the second point, you draw a line again from the original position of Point D to the final position of that point. I label this line "Line of translation #2." If these two lines of translation remain parallel from the original location to the final location, a pure translational behavior has occurred. These lines of translation can occur at any orientation in space, including in three dimensions.

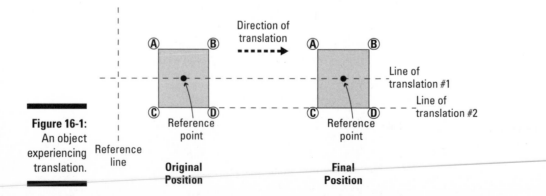

Figure 16-1:
An object experiencing translation.

One of the ways you can prove whether two lines of translation are parallel is by using unit vectors. Parallel lines have identical unit vectors.

Rotational equilibrium

Any behavior that results in a change in the orientation of an object without causing it to translate is considered to be a *rotational* behavior.

Rotation is defined with respect to an axis. In Chapter 12, I explain that a moment is a rotational effect that acts about an axis of rotation. In two dimensions, you don't actually see this axis of rotation but rather the location where the axis of rotation intersects the *xy* plane. In this chapter, I identify that point as the "reference point."

Be sure not to confuse rotation with revolving or orbiting, which implies a motion around or about another object. The moon orbits the Earth, and the Earth revolves around the sun. However, the Earth spins, or rotates, about its own polar axis. Consider the object of Figure 16-2, which illustrates the difference between rotating (Figure 16-2a) and orbiting (Figure 16-2b).

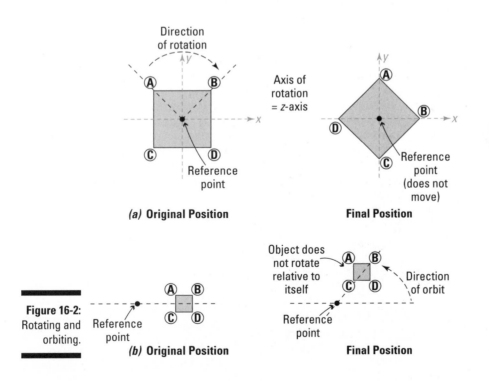

Figure 16-2:
Rotating and orbiting.

To determine whether an object is experiencing rotation, consider the orientation of two points on the object, relative to the axis of rotation (the reference point). For this example, I arbitrarily choose Point A and Point B on the boundary of the object. If the reference point remains unchanged, and Point A and Point B experience the same change in angle with respect to the reference point (the *z*-axis in this two-dimensional example), you can conclude that the object has experienced a pure rotational behavior. Figure 16-2b shows an object experiencing an orbit effect. In this case, the orientation of the object with respect to itself remains unchanged. That is, the corner A doesn't change, despite its change in orbital location.

The axis of rotation (see Chapter 12) doesn't have to be parallel to a Cartesian axis, though in many two-dimensional problems in the *xy* Cartesian plane, the rotation occurs about the *z*-axis. For three-dimensional problems, the axis of rotation can actually have any orientation in space. Again, you can use unit vectors to help you describe this orientation.

Rotation is always caused by the application of an applied moment, couple, or eccentric load (see Chapter 12 if you need more information). If an object is subjected to unbalanced moments, couples, or eccentric loads, that object experiences a rotation. Thus, if all the rotational behaviors are balanced, an object is in *rotational equilibrium,* and the net effect of the moments (or the *resultant*) of all the rotational behaviors must also be zero.

In the real world, you can observe an endless combination of translational and rotational effects. Objects can be *displacing* (translating) and, at the same time, *spinning* (rotating). *Rolling* is actually a combination of translation and rotation. For example, a tire on a moving car experiences rolling. The tire itself is rotating about its center point (the axle of the car), but at the same time, the center of the tire is moving in a straight line in the direction of the traveling car (assuming, of course, that the car is traveling on a straight and level stretch of road).

In order for an object to be in total equilibrium, it must be balanced for all translational behaviors at the same time that it's balanced for the rotational behaviors. If either one of these behaviors isn't balanced, the object can't be considered to be in equilibrium.

Looking for Equilibrium with Newton's Laws

The basic equations of statics and equilibrium are founded in the principles of Newtonian mechanics, developed by Sir Isaac Newton in the 17th century. Newton's three laws of motion help describe the way forces and objects interact, and ultimately provide the basic equations of equilibrium.

Reviewing Newton's laws of motion

Newton's laws provide a solid foundation for dealing with objects subjected to forces (also known as the study of mechanics). Perhaps the most famous of his explanations are contained in his three fundamental laws of motion:

- **Newton's first law:** Newton's *first law of motion,* sometimes referred to as the *law of inertia,* states that an object at rest tends to stay at rest until acted upon by an unbalanced force. Likewise, an object in motion stays in motion with the same speed and in the same direction until acted upon by an unbalanced force.

- **Newton's second law:** Newton's *second law of motion* states that when a force acts upon an object that has mass, a corresponding acceleration is produced. This idea led to one of the most popular formulas in all of physics:

   $\vec{F} = m \cdot \vec{a}$, where **F** represents the unbalanced applied force (the resultant), *m* represents the mass of the object, and **a** represents the resulting acceleration due to those applied forces.

- **Newton's third law:** Newton's *third law of motion* is one of the most commonly known physics expressions. It states that for every action, there is an equal and opposite reaction.

Newton's third law helps you in creating free-body diagrams. Drawing F.B.D.s of parts of objects or systems is often more convenient than drawing entire systems. Newton's third law allows you to replace parts of a complete picture with reaction forces. I show you exactly how this is done in Chapters 17 and 18. For now, check out the example in Figure 16-3, where you notice a person pushing on a crate without moving it.

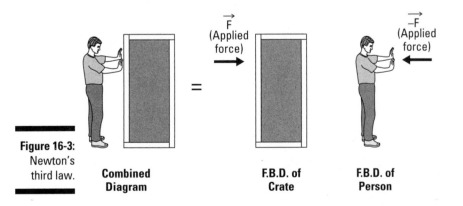

Figure 16-3:
Newton's
third law.

| Combined Diagram | F.B.D. of Crate | F.B.D. of Person |

How a falling apple changed the world

Sir Isaac Newton (1642–1727) was an English astronomer, physicist, mathematician, and philosopher who is credited with developing the basic equations of modern physics. Legend has it that as he was walking through his orchard (or sitting under an apple tree, depending on which version of the story you hear), he observed an apple falling from a tree (or it hit him on the head, again depending on the version of the story). From this simple event, Newton formulated several basic ideas that completely redefined the way that physicists, scientists, and engineers look at the world around them. He published these thoughts, and others, in a document known as *Philosophiæ*

Naturalis Principia Mathematica, which was first published around 1687.

For almost 200 years (until the late 19th century), these basic rules were indisputable. It wasn't until the development of several very advanced theories — such as quantum mechanics theory by Max Planck around 1900 and the general theory of relativity by Albert Einstein around 1905 — that the Newtonian laws began to break down on the atomic and cosmic scales. However, in day-to-day applications, even to this day, Newtonian mechanics remain good tools for working with basic mechanics and physics problems, such as the ones discussed in this book.

If you examine the free-body diagram of just the crate, you notice an applied force **F** from the person pushing on the crate. Similarly, if you remove the crate from the picture, that person, in reality, is still standing stationary. Without the crate, the force the person applies would cause that person to move in the direction of his applied force. However, because the man isn't actually moving in reality, there must be an applied force from the box onto the person, which is helping to keep the person in his position. The force from the box onto the man has the same magnitude and line of action, but in the opposite direction, or **–F**. The net effect of these two forces is then **F** + (**–F**) = 0. There is no net unbalanced force between the man and the crate, so both are said to be in a state of translational equilibrium.

The scalar equations that make it happen: The big three

As I discuss in Chapter 1, one of the largest assumptions in statics comes from the very definition of the word *static,* which means "constant" or "unchanging." If an object's position and orientation are unchanging, its velocity and acceleration should clearly both be zero. If **a** = 0, Newton's second law states that **F** = $m(0)$ = 0. Thus, for an object to remain at rest, or in equilibrium), Newton's second law requires that the resultant of all the forces and all of the moments acting on the object must be zero.

In two dimensions

In two-dimensional problems, in order to fully define translational equilibrium, you need to consider translational behaviors in at least two nonparallel directions, which are often taken to be perpendicular to each other and in the same direction as the Cartesian axes of your applied coordinate systems (typically, in the x- and y-directions).

Now, if you want to study the behavior of a force in the x-direction only, you need to only look at the x-components, and verify that the resultant of all the x-component vectors is zero. In equation form, this setup means

$$\sum \overrightarrow{F_x} = 0$$

Likewise, you must also consider at least one other direction (not parallel to the Cartesian x-axis) and verify that translation equilibrium occurs in that direction as well. In equation form, this requirement means

$$\sum \overrightarrow{F_y} = 0$$

After defining translational equilibrium, you also need to verify that there are no rotational behaviors, or *moments*. In order to ensure rotational equilibrium, the resultant of all moments on the object must be balanced. In equation form, this is expressed as

$$\sum \overrightarrow{M_z} = 0$$

In three dimensions

You can apply the same logic you used for two-dimensional problems when solving three-dimensional problems. However, you need to

✔ Evaluate translational equilibrium in at least three nonparallel directions (usually taken as the Cartesian x-, y-, and z-directions)

✔ Consider rotational equilibrium by considering the moment affects about three nonparallel axes of rotation

To verify equilibrium in three dimensions, you have a total of six equations at your disposal — three for translation and three for rotation:

✔ **For translation:**

$$\sum \overrightarrow{F_x} = 0$$

$$\sum \overrightarrow{F_y} = 0$$

$$\sum \overrightarrow{F_z} = 0$$

✔ **For rotation:**

$$\sum \overrightarrow{M_x} = 0$$

$$\sum \overrightarrow{M_y} = 0$$

$$\sum \overrightarrow{M_z} = 0$$

Identifying Improper Constraints: When Equilibrium Equations Are Insufficient

Sometimes you can determine whether a system is in equilibrium by being able to identify unique situations such as concurrent force systems and parallel force systems. I explain both of these systems in the following sections.

Concurrent force systems

A *concurrent force system* applies to objects that have been subjected to a system of forces whose lines of action all act through a common point. In Figure 16-4, forces F_1, F_2, and F_3 all act concurrently through Point O.

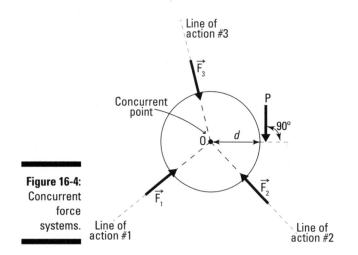

Figure 16-4: Concurrent force systems.

Suppose that those forces represent the support reactions for the object. Because these forces all have x- and y-components, those three forces may mathematically result in translational equilibrium — that is, those three forces may all balance each other.

However, if a new force **P** is applied somewhere else on the object (such as a perpendicular distance d from the concurrent point), an interesting phenomenon occurs: The reaction forces — $\mathbf{F_1}$, $\mathbf{F_2}$, and $\mathbf{F_3}$ — change values depending on the force vector **P** in order to maintain translational equilibrium. But if you calculate the equivalent moment of all the forces at the concurrent point, O, you have

$$\left|\overrightarrow{M_o}\right| = \left|\overrightarrow{F_1}\right| \cdot 0 + \left|\overrightarrow{F_2}\right| \cdot 0 + \left|\overrightarrow{F_3}\right| \cdot 0 + \left|\overrightarrow{P}\right| \cdot d = \left|\overrightarrow{P}\right| \cdot d$$

From this equation, if d is nonzero (in other words, the load is at some point other than the concurrent point) and if the magnitude of the load, $|P|$, is also nonzero, the equivalent moment at the concurrent point can never be equal to zero. Physically, this fact means that any single force you apply to the object of a concurrent force system at a point other than the concurrent point results in a rotational behavior.

In order for rotational equilibrium to occur, the equivalent moment at all points must be equal to zero. If you have a rotational behavior, you can't have rotational equilibrium.

So, by simply examining the lines of action of the support reactions, you can determine whether a given problem is a concurrent force problem, which in turn gives you insight into the state of equilibrium of the problem. For this example, there is no rotational equilibrium. If more than one force is applied, rotational equilibrium is still possible.

Parallel force systems

Like the name implies, a *parallel force system* applies to objects where all applied forces have parallel lines of action. In Figure 16-5, forces $\mathbf{F_1}$, $\mathbf{F_2}$, and $\mathbf{F_3}$ are all parallel to each other. The magnitude and sense of these force vectors may vary, but their lines of action must remain parallel.

Based on their magnitudes, the three forces — $\mathbf{F_1}$, $\mathbf{F_2}$, and $\mathbf{F_3}$ — may feasibly balance each other and result in translational equilibrium. If a new force **P** is applied to the system, and if the line of action of the new force is parallel to the original three forces, translational equilibrium may still be maintained.

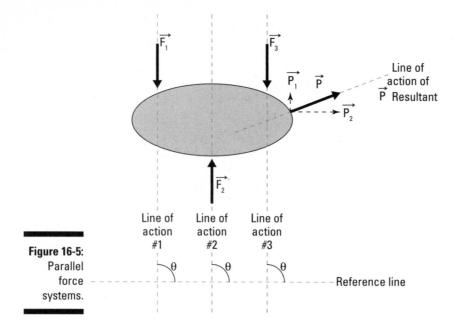

Figure 16-5:
Parallel
force
systems.

However, if the new force **P** is *not* parallel to the original three forces, you need to do some more investigating. If you compute the components of vector **P** such that one of the components (P_1) is parallel to the forces F_1, F_2, and F_3, equilibrium may be achieved in the direction of those forces. But remember that for two-dimensional problems, you must always include both components in your substitution for the force, **P** — and that second component is what causes the problem. If component P_1 is parallel to the original forces, component P_2 must mathematically be equal to zero, and equilibrium in two unique directions can be guaranteed. However, if the force **P** is oriented in any direction other than parallel to the original forces, the component P_2 can never be zero. In this case, the translation in the parallel direction may be balanced, but the translation in the direction of the second component will never be balanced.

In order for translational equilibrium to be satisfied, the net translational behavior in two unique directions (three directions for three-dimensional problems) must be zero. If one direction is unbalanced, equilibrium can't be satisfied.

Chapter 17

Taking a Closer Look at Two-Dimensional Equilibrium: Scalar Methods

. . . .

In This Chapter

▶ Solving for reactions

▶ Summing forces and moments

▶ Identifying alternative locations for moment equations

. . . .

The equilibrium equations I present in Chapter 16 give you the tools you need to begin studying the effects of behaviors on structures. You're well on your way to determining support reactions, calculating internal forces, and solving specific application problems with either scalar or vector techniques. For two-dimensional problems, scalar solution techniques are much more efficient, and that's what I show you in this chapter. But never fear — I show you the vector methods (which you almost always need for three-dimensional problems) in Chapter 18. In this chapter, I show you how to use your free-body diagrams (F.B.D.s — see Part IV) to determine the magnitude and sense of unknown support reactions. I start the chapter by outlining the three basic steps that you follow when working a scalar statics problem and then I show you how to create the translational and rotational equilibrium equations in two-dimensional situations. Finally, I highlight a few considerations for selecting points for moment equilibrium equations.

Tackling Two-Dimensional Statics Problems in Three Basic Steps

As you start to analyze a system or object, keep in mind that you're always using the concepts of static equilibrium to determine the unknown forces and behaviors. Although some problems (such as those defined in Part VI) may require specific techniques to unravel them, you quickly discover that the vast majority of statics problems begin with the same basic steps:

1. **Draw a free-body diagram of the object of interest.**

   Construct your F.B.D. by including all information about the applied external forces, any revealed internal forces, the self weight, and unknown support reactions.

   Even if you don't know the magnitude or sense of a particular action at this point, go ahead and include it on the diagram. Just make sure that the action is located at its proper point of application and oriented with its proper line of action, and give the load a label.

2. **Write the equations of equilibrium for your free-body diagram.**

   The number of equations you have to write depends on the number of dimensions of your free-body diagram:

   • **For two-dimensional problems:** If you've constructed a two-dimensional F.B.D., you write three equations. You need two translational equations (sum of forces) and a rotational equation (sum of moments).

   • **For three-dimensional problems:** If you've constructed a three-dimensional F.B.D., you need six scalar equations (three translational equations and three rotational equations).

   In this step, your primary goal is to determine the magnitude of as many of the any unknown support reaction forces as possible.

3. **Calculate any necessary internal forces.**

   After you've calculated as many of the external reactions as possible, your next step is usually to determine internal forces and other more specific information about the object(s) you're working with. Many of the problems you encounter within statics have specific solution techniques once the support reactions have been determined. I highlight these different types of problems in the chapters of Part VI.

In the coming sections, I show you the techniques for computing the support reactions.

Calculating Support Reactions with Two-Dimensional Equilibrium Equations

When you know the steps (see the preceding section), you're ready to put Mr. Newton's equations to work for you.

Consider the structure of Figure 17-1a, which is loaded with a 0.5-kip-per-inch uniformly distributed load from Point A to Point B, a concentrated load of 10 kip at Point B, and a concentrated moment of 40 kip-inches at Point C. The structure is supported by a pinned support at Point A and a roller support on a 30-degree incline ramp at Point D. In the following sections, I show you how to determine the magnitudes of the external support reactions using this example.

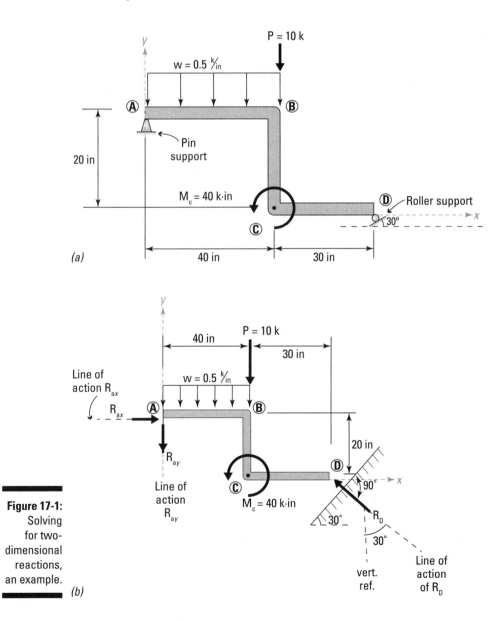

Figure 17-1:
Solving for two-dimensional reactions, an example.

(a)

(b)

First things first: Creating the F.B.D.

Your first step is to create the system F.B.D. by using the tools in Part III. Make sure to include the following four types of forces on your diagram:

✔ **External forces:** In the example of Figure 17-1, they're actually provided for you. Just copy the distributed load, the concentrated load, and the moment to their locations on the F.B.D. as shown in Figure 17-1b.

✔ **Internal forces:** For this F.B.D., you have no internal forces to include because you don't have to actually cut the structure.

✔ **Self weight:** Neither the mass nor the self weight is given, so you can assume that the self weight is negligible in this problem. This example has no self weight forces.

✔ **Support reactions:** In this example, the support at Point A is a pinned support, which means that you need to include at least two unknown support reactions. Even though you don't know the sense or the magnitude at this time, you can work around this problem. (I show you how in the "Solving for the unknown reactions" section later in this chapter.)

The roller support at Point D is a bit more complicated. *Remember:* A roller support must have a single reaction force that is normal (or perpendicular) to the plane of the support. In this example, the support is sloped at 30 degrees from the horizontal, so the line of action of this reaction must be perpendicular to the plane of the support.

Writing the equilibrium equations

In Chapter 16, I introduce you to the equations that you need to develop in order to establish equilibrium conditions. For a two-dimensional problem, you need to write two translational equilibrium equations (sum of forces) and one rotational equilibrium equation (sum of moments):

$$\overset{+}{\rightarrow}\sum \overrightarrow{F_x} = 0 \qquad +\!\!\downarrow \sum \overrightarrow{F_y} = 0 \qquad \overset{\curvearrowleft}{+}\sum \overline{M} = 0$$

Adding helpful notation

When using the equations of equilibrium, notice that the equations contain vector expressions. However, you can actually cheat a little if you consistently assume a vector direction. If two vectors, F_1 and F_2, are both acting in the positive Cartesian x-direction, you can write their vector expressions in terms of their scalar magnitudes and their Cartesian unit vectors:

$$\overrightarrow{F_1} = \left|\overrightarrow{F_1}\right| \cdot \hat{i}$$

$$\overrightarrow{F_2} = \left|\overrightarrow{F_2}\right| \cdot \hat{i}$$

The equilibrium equation in the x-direction is then written as

$$\sum \vec{F}_x = 0 \rightarrow \vec{F}_1 + \vec{F}_2 = 0 \rightarrow \left(\left|\vec{F}_1\right| + \left|\vec{F}_2\right|\right) \cdot \hat{i} = 0 \rightarrow \sum \left|\vec{F}_x\right| = \left|\vec{F}_1\right| + \left|\vec{F}_2\right| = 0$$

Notice how the unit vector doesn't actually affect the outcome in the expression. At the end, you're left with a scalar expression that involves only the magnitudes of the forces acting in that direction.

I also recommend including a bit of notation before the expression to help you remember to be consistent with the directions that you assume to be positive:

$$\xrightarrow{+} \sum \vec{F}_x = 0$$

In this case, I like to add an arrow indicating the direction that I assume to be positive. This arrow serves as a constant reminder as I establish the equations of equilibrium.

It doesn't matter which way you assume is positive when writing the equilibrium equations — just pick a direction as positive and then be consistent with it when you start writing the equations.

Similarly, the moment vector equation can also be simplified to scalar expressions if you choose a consistent direction for the axis of rotation. You can also add a similar reminder before each of your moment equations:

$$\circlearrowleft + \sum \left|\overline{M}\right| = 0$$

Summing forces first: Writing two translational equilibrium equations

When you're solving for reaction forces, it doesn't really matter which equilibrium equation you write first, so for this example I start with summing forces in the positive x-direction. To write this equilibrium expression, you must include every component of a force (see Chapter 8) that acts in the positive x-direction. For example, consider the reaction at Figure 17-1's Point D, shown in Figure 17-2.

On the F.B.D., I assume that the force is acting up and to the left. Based on the assumed direction of R_D, the corresponding x-component R_{Dx} must be acting to the left, and the y-component R_{Dy} is acting upward. You must get these directions correct, with respect to your assumed direction for R_D. A mistake in the directions of force components will often prevent you from getting a final correct answer.

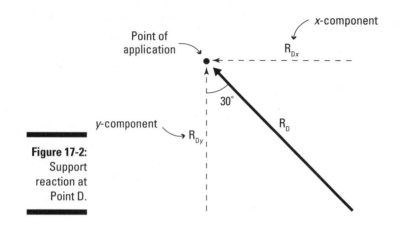

Figure 17-2:
Support
reaction at
Point D.

The scalar equilibrium equation for translation in the x-direction is

$$\overset{+}{\to}\sum\left|\vec{F}_x\right| = 0 \to + R_{Ax} - R_{Dx} = 0 \to + R_{Ax} - R_D \sin(30°) = 0$$

In this equation, I've indicated that the positive direction is in the direction of the positive x-axis (or to the right). R_{Ax} is entered into the equation as a positive value because it is acting in the same direction as the assumed positive direction. Likewise, the x-component of R_D is negative because it acts in the opposite direction. This is one of the expressions required for translational equilibrium. At this point, you can't actually solve this equation because there are two unknown values in this equation. However, you have created a relationship between R_{Ax} and R_D that satisfies Newton's laws of equilibrium.

Next, you can write the translational equilibrium in the y-direction:

$$\sum\left|\vec{F}_y\right| = 0 \to -R_{Ay} + R_{Dy} - 10 \text{ k} - (0.5 \text{ k in})\cdot(40 \text{ in}) = 0$$
$$\to -R_{Ay} + R_D \cos(30°) = 30 \text{ k}$$

In this equation, the vertical reaction R_{Ay} is entered as a negative value because a positive direction was assumed upward, and this force was assumed to be acting downward. R_{Dy} is positive because the vertical component of the assumed R_D is acting upwards. Finally, all the external forces are included. The 10-kip concentrated force is negative because it is acting downward, and the resultant of the distributed load (0.5 k/in) · (40 in) is also acting downward.

Notice, however, that I haven't included the concentrated moment in either of these translation equations. Concentrated moments show up only in the rotational equilibrium equation because they're rotational behaviors — they don't show up in the translation equations. However, as you see in the following section, both concentrated and distributed forces show up in the moment equilibrium equations.

Summing moments: Writing the rotational equilibrium equation

The third equation that you need to write is the rotational equilibrium equation. This equation behaves a bit differently because, unlike the sum of forces equations that you just wrote, the moment equation requires you to pick a specific point about which you calculate the equivalent resultant moments for the system.

For the sake of this example, I arbitrarily choose Point B from Figure 17-1 as the location and see what happens. (If you want to see an even easier way to solve this problem, skip ahead to the "Choosing a Better Place to Sum Moments" section later in the chapter.) You calculate the equivalent moment of each of the actions on the structure about Point B as shown in Figure 17-3.

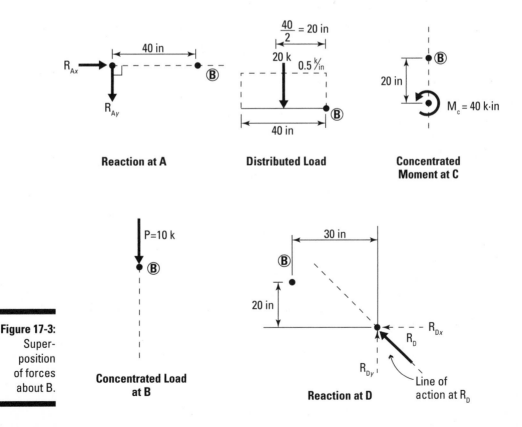

Figure 17-3: Superposition of forces about B.

Reaction at A

Distributed Load

Concentrated Moment at C

Concentrated Load at B

Reaction at D

Assuming a counterclockwise moment to be positive, the rotational equilibrium equation about Point B is

$$\circlearrowleft + \sum |\overline{M_z}|^B = 0 \rightarrow \cancel{R_{Ax}(0)} + R_{Ay}(40 \text{ in}) + (0.5 \text{ k/in})(40 \text{ in})\left(\frac{40 \text{ in}}{2}\right)$$

$$+40 \text{ k} \cdot \text{ in} + \cancel{10\text{ k}(0)} + R_{Dy}(30 \text{ in}) - R_{Dx}(20 \text{ in}) = 0$$

$$\rightarrow +R_{Ay}(40 \text{ in}) + R_D \cos(30)(30 \text{ in}) - R_D \sin(30)(20 \text{ in}) = -440 \cdot \text{in}$$

Each term in the moment equation is the equivalent moment of an action about Point B (which is the point you're summing moments about). Each force (or components) is included in the expression and includes the perpendicular distance to Point B. If it causes a rotation in the opposite direction, you include a negative sign before that term. A concentrated moment (40 kip per inch, in this example) is included in the equilibrium equation as a single value. It is added (or subtracted depending on its direction of rotation) to the equation directly.

If you have a concentrated moment, this value does not require an associated distance. A moment is a free vector, and its equivalent moment at any location is the same as the original moment. The sign before this term in the equation is based on the assumed positive direction of rotation.

Simplifying, you can derive the final rotational equilibrium equation with respect to Point B:

$$+R_{Ay}(40 \text{ in}) + R_D\left[\cos(30)(30 \text{ in}) - \sin(30)(20 \text{ in})\right] = -440 \text{ k} \cdot \text{in}$$

$$\rightarrow R_{Ay}(40 \text{ in}) + R_D(15.98 \text{ in}) = -440 \text{ k} \cdot \text{in}$$

Solving for the unknown reactions

By examining each of the equilibrium equations, you can see that each of the three unknown reactions appears in at least one of the equilibrium equations (and in the case of R_D, multiple appearances):

- ✔ $+R_{Ax} - R_D \sin(30°) = 0,$
- ✔ $-R_{Ay} + R_D \cos(30°) = 30 \text{ k}$, and
- ✔ $R_{Ay}(40 \text{ in}) + R_D(15.98 \text{ in}) = -440 \text{ k} \cdot \text{in}$

With a few basic algebra skills, simultaneous solution of these equations produces the values of the magnitudes for R_D, R_{Ax}, and R_{Ay}. Of course, the math isn't exactly friendly, but the statics is complete, and you're now left with solvable equations that produce the following reactions:

$$R_D = +15.0 \text{ k} = 15.0 \text{ k} \searrow$$
$$R_{Ay} = -17.0 \text{ k} = 17.0 \text{ k} \uparrow$$
$$R_{Ax} = +7.50 \text{ k} = 7.50 \text{ k} \rightarrow$$

After the algebra is complete, if you find that a particular numerical result has a positive value, you know that the direction you assumed on the F.B.D. was correct for that load. Notice that R_{Ay} had a value of –17.0 kip, which tells you that instead of acting downward as you originally assumed, it's actually acting in the opposite direction (or upward).

As long as you're consistent in assigning directions for all your scalar values in the equations, the signs will tell you whether your assumptions are correct.

When you work with scalar equilibrium problems, if you make even a single mistake with one sign or assumed direction, you're probably done for. It's not because scalar methods are overly that difficult — it's just human nature to make simple sign error mistakes from time to time.

Choosing a Better Place to Sum Moments

When you're writing the rotational scalar equilibrium equation, you can choose any point about which you sum moments, even a point not physically on the structure. Referring to the Figure 17-1 example in the preceding sections, if you were to sum moments about Point A rather than Point B, you would eliminate the reactions at Point A from the equilibrium equation altogether, because at this summation location, the perpendicular distances from the point of interest to the reaction forces at Point A are zero.

$$\circlearrowleft + \sum \left|\overline{M_z}\right|^A = 0 \rightarrow \cancel{R_{Ax}(0)} + \cancel{R_{Ay}(0)} - (0.5 \text{ k/in})(40 \text{ in})\left(\frac{40 \text{ in}}{2}\right)$$
$$\rightarrow +40 \text{ k} \cdot \text{in} - 10 \text{ k}(40 \text{ in}) + R_{Dy}(70 \text{ in}) - R_{Dx}(20 \text{ in}) = 0$$
$$\rightarrow R_D \left[\cos(30)(70 \text{ in}) - \sin(30)(20 \text{ in})\right] = +760 \text{ k} \cdot \text{in}$$

Now this equation only has one unknown, R_D, and you can solve for it directly, making the math significantly easier than summing moments at Point B.

$$R_D = +15.0 \text{ k} = 15.0 \text{ k} \nwarrow$$

After you have the value and sense for R_D, you can then compute the reactions R_{Ax} and R_{Ay} from the translational equilibrium equations.

By carefully selecting the points about which you sum moments, you can greatly simplify your equation writing. When you're looking for possible locations for rotational equilibrium points, you may want to consider the following:

✔ **Pinned support locations and internal hinges:** Pins and hinges always have two unknown forces associated with them.

✔ **Locations where lines of action of multiple unknown forces intersect:** This point is known as an *instantaneous center* or a *point of concurrency*.

✔ **Locations that eliminate troublesome forces with inconveniently oriented lines of action:** As the last example shows, summing moments at Point D eliminates some nasty trig calculations from your equations. If you're not sure how to handle an unknown force, sum moments at that point so that you can eliminate it from the rotational equilibrium equation altogether. But remember, just because you eliminate it from the sum of moments equation doesn't mean you can eliminate it from the sum of forces equation.

For example, if you had chosen to sum moments at Point D:

$$\circlearrowleft + \sum \left|\overline{M_z}\right|^D = 0 \rightarrow -R_{Ax}(20 \text{ in}) + R_{Ay}(70 \text{ in}) + (0.5 \text{ k in})(40 \text{ in})\left(\frac{40 \text{ in}}{2} + 30 \text{ in}\right)$$

$$\rightarrow +40 \text{ k} \cdot \text{in} + (10 \text{ k})(30 \text{ in}) + R_{Dy}(0) - R_{Dx}(0) = 0$$

$$\rightarrow -R_{Ax}(20 \text{ in}) + R_{Ay}(70 \text{ in}) = -1,340 \text{ k} \cdot \text{in}$$

Notice how choosing Point D eliminates all the trigonometry terms that showed up in the other equations. However, the problem with summing moments at Point D, however, is that terms for R_{Ax} and R_{Ay} both appear again, which requires solving simultaneous equations again. Regardless, the final solution is the same regardless of which moment summation point you choose.

Chapter 18

Getting Better Acquainted with Three-Dimensional Equilibrium: Vector Methods

. . . .

In This Chapter

▶ Solving for reactions in three-dimensional problems

▶ Explaining equilibrium in three dimensions

▶ Working with three-dimensional vectors in equilibrium equations

. . . .

Statics problems in three dimensions are some of the most intimidating types of problems you encounter. You can easily get lost in all the dimensions, coordinates, and angles in three dimensions. A single force in three dimensions is completely capable of producing moments about three different Cartesian axes (see Chapter 5). And determining whether a rotation is positive or negative by the scalar methods I show in Chapter 17 can be a real challenge (though not impossible with the right-hand rule I describe in Chapter 12). If you choose to work with vectors, a lot of those difficulties are automatically handled for you. And after you've become an old pro at turning forces and distances into vectors, why not let vectors do the work for you?

In this chapter, I show you how to apply all those vector calculations to the free-body diagrams (F.B.D.s) I cover earlier in the book. Then I show you how to write the equations of equilibrium in vector form to solve for unknown support reactions.

Finding a Starting Point

The solution method for solving three-dimensional statics problems is very similar to the solution methods for two-dimensional situations that I describe in Chapter 17. Although it requires a bit more work upfront, the actual completion of the equilibrium equations is much easier from a statics point of view; the math, on the other hand, can be a bit lengthy. To solve for support reactions in three-dimensional problems, you follow a similar set of steps to the two-dimensional scalar methods, with a few minor exceptions and additions.

Following is a basic summary of the steps required to find the support reactions for a three-dimensional problem:

1. **Draw a free-body diagram of the object of interest.**

   You include all concentrated forces, distributed forces, and moments on your F.B.D. Be sure to include the necessary information for describing the lines of action of your forces and the axes of rotations for your concentrated moments. You need to construct unit vectors in this process, so be sure to include any necessary information to construct position vectors in order help you define these important directions. Also, remember that you must provide coordinate data and angles.

2. **Write the equations of three-dimensional equilibrium.**

   The number of equilibrium equations that are required to solve a three-dimensional problem depends on the method you use to solve the problem:

   • **Scalar methods:** If you plan to perform scalar equilibrium calculations on three-dimensional problems, you need to work with a total of six equilibrium equations per F.B.D. — three scalar equations for translational equilibrium and another three scalar equations for rotational equilibrium. (See Chapter 16 for these six equations.)

   Working a scalar method in three dimensions is actually no different than it is in two dimensions. The process is just a little trickier in three dimensions because you need to keep track of the directions (whether they're positive or negative) by hand, just like you do with the two-dimensional problems of Chapter 17. However, in three dimensions, a force can create a rotation about any of the three Cartesian axes, and keeping track of those signs and magnitudes can be a bit confusing.

   • **Vector methods:** For vector method solutions, the process is actually a lot more straightforward. Instead of needing six scalar equations, you need only two vector equations to establish equilibrium:

   $$\sum \vec{F} = 0 \qquad\qquad \sum \vec{M} = 0$$

   Although you're working with fewer equations, you have to do a few more calculations upfront — you need to create vectors out of every force and moment on your free-body diagram. (I describe many of those methods earlier in the book, so if you need a refresher, look to Chapters 5, 7, and 8.)

You use these equations to determine as many of the external reactions as possible. If your F.B.D. has no more than six unknown reactions, these two vector equations can help you calculate the magnitudes of all of them. If your F.B.D. has more than six reactions, you may not be able to actually determine their magnitudes, but you can to write expressions that relate them to each other.

3. Calculate necessary internal forces (see Part VI for these techniques).

Seeing Equilibrium within Vector Notation

In Chapter 16, I define *equilibrium* for an object as the state that occurs when all translational and rotational behaviors are balanced at the same time, and in the preceding section, I show you the two basic vector equations needed to define equilibrium. In this section, I explain each of these equations in more detail and show you how the vector equations actually contain the scalar equations in them automatically.

Equilibrium in translational behaviors

For an object to be in translational equilibrium, the net effect of all forces must be in balance. In statics terms, this setup means that the resultant of all translational behaviors (forces) must be equal to zero.

In Chapter 16, I mention that it takes a minimum of three translational directions to establish equilibrium for a three-dimensional object. So, if you arbitrarily choose the Cartesian x-, y-, and z-axes to represent your three directions, the resultant force **F** acting on an object can be expressed as

$$\sum \vec{F} = \sum \vec{F_x} + \sum \vec{F_y} + \sum \vec{F_z} = \sum \left| \vec{F_x} \right| \cdot \hat{i} + \sum \left| \vec{F_y} \right| \cdot \hat{j} + \sum \left| \vec{F_z} \right| \cdot \hat{k}$$

If an object is said to be in equilibrium, the resultant vector **F** must take the following form:

$$\sum \vec{F} = 0\hat{i} + 0\hat{j} + 0\hat{k}$$

In order for these two expressions to be equal, the following expressions must be true:

$$\sum \left| \vec{F_x} \right| = 0 \qquad \sum \left| \vec{F_y} \right| = 0 \qquad \sum \left| \vec{F_z} \right| = 0$$

That means that the sum of forces in each of the three Cartesian directions must be equal to zero. Notice that these three equations are the same translational equilibrium equations I mention in Chapter 16.

Rotational components

In order for an object to be in a state of equilibrium, all the rotational behaviors (or the moments) must also be balanced — that is, the resultant of all rotational effects must also be zero. To establish rotational equilibrium, you need to ensure that the rotational effects about three different non-coplanar axes are also balanced. Following the logic for translational equilibrium in the preceding section, you can write a similar expression for the resultant moment vector **M** acting on an object as the following:

$$\sum \overline{\mathbf{M}} = \sum \overrightarrow{\mathbf{M}_x} + \sum \overrightarrow{\mathbf{M}_y} + \sum \overrightarrow{\mathbf{M}_z} = \sum \left|\overrightarrow{\mathbf{M}_x}\right| \cdot \hat{i} + \sum \left|\overrightarrow{\mathbf{M}_y}\right| \cdot \hat{j} + \sum \left|\overrightarrow{\mathbf{M}_z}\right| \cdot \hat{k}$$

For an object to be in rotational equilibrium, the resultant moment **M** must be given by

$$\sum \overline{\mathbf{M}} = 0\hat{i} + 0\hat{j} + 0\hat{k}$$

This resultant implies that the scalar components of the moment about any three axes must be equal to zero to ensure rotational equilibrium:

$$\sum \left|\overrightarrow{\mathbf{M}_x}\right| = 0 \qquad \sum \left|\overrightarrow{\mathbf{M}_y}\right| = 0 \qquad \sum \left|\overrightarrow{\mathbf{M}_z}\right| = 0$$

Notice that these three equations are the same rotational equations that must be satisfied using the scalar methods.

Figuring Support Reactions with Three-Dimensional Equilibrium Equations

As with two-dimensional equilibrium problems, you always want to try to compute all of the magnitudes of the unknown support reactions — or as many of the magnitudes as possible — after drawing the F.B.D. of the system. Three-dimensional problems are no different.

In Figure 18-1, a uniform 2-meter-x-3-meter plank with a mass of 10 kilograms is supported by a three-dimensional roller at Point A and a three-dimensional pinned support (or ball and socket) at Point D. Both Points B and C are tied with cables to a point on the wall at Point E. A concentrated point load of 100 Newton is applied on side *AB* at an angle of 30 degrees below the horizontal, and parallel to the *yz* plane. A moment of 300 Newton-meters is applied along edge *BC.* In the following sections, I walk you through the basic process for computing the magnitudes of the reactions for a three-dimensional statics problem. But first, as always, you need to create a proper free-body diagram first.

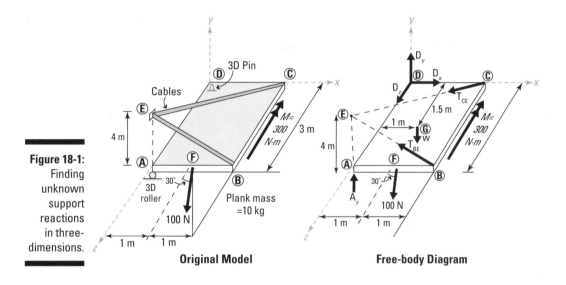

Figure 18-1:
Finding
unknown
support
reactions
in three-
dimensions.

Original Model   **Free-body Diagram**

Establishing the F.B.D.

With any statics problem, regardless of whether it's two-dimensional or three-dimensional, your first step in the solution process is always to sketch a free-body diagram of the forces and moments that are acting on the object. Then you need to find as many of the unknown support reactions as possible.

Sketching the loads on the F.B.D.

In the example of Figure 18-1, if you draw an *isolation box* (which lets you zoom in on a specific feature of a larger object) around the plank by cutting the two cables, you can expose the support reactions that you're seeking. (See Chapter 14 for more on using isolation boxes.)

✔ **External forces:** Typically, you know the external forces acting on a system. The external forces for this problem are the 100-Newton concentrated point load alongside *AB* and the 300-Newton-meter concentrated moment along edge *BC*.

✔ **Internal forces:** Internal forces appear anytime a physical structure is cut. In this example, you're cutting both of the cables BE and CE in order to isolate the F.B.D. of the plank. You know that cables are axial members only, so you need to include a force for cable BE, T_{BE}, along the line from Point B to Point E. Similarly, you include another force for cable CE, T_{CE}, acting along a line from Point C to Point E. Even though you don't know the magnitude of these forces, you must include them on the free-body diagram.

✔ **Support reactions:** The external support reactions in this example are a three-dimensional roller at Point A, which has a single point load A_y acting upward in the positive *y*-direction, and a three-dimensional pinned support at Point D that has three component forces (D_x, D_y, and D_z) acting parallel to each of the Cartesian axes.

I should point out that in this example, you can also combine these scalar reactions into a single resultant (see Chapter 7) that represents all of the concentrated forces for the support reaction. For $\mathbf{R_D}$, the reaction resultant vector at Point D:

$$\overrightarrow{R_D} = D_x \cdot \hat{i} + D_y \cdot \hat{j} + D_z \cdot \hat{k}$$

where D_x, D_y, and D_z represent the scalar magnitudes of the components in the *x*-, *y*-, and *z*-directions respectively. Similarly, for the reaction at Point A:

$$\overrightarrow{R_A} = A_x \cdot \hat{i} + A_y \cdot \hat{j} + A_z \cdot \hat{k}$$

However, because the support at Point A is a three-dimensional roller, you automatically know that $A_x = 0$, and $A_z = 0$, so the reaction at Point A can be simplified to

$$\overrightarrow{R_A} = A_y \cdot \hat{j}$$

✔ **Self weight:** In this problem, you also know that the plank has a mass of 10 kilograms, which means that you need to include the self weight on your free-body diagram. Compute the weight as $W = mg = (10 \text{ kg}) \cdot (9.81 \text{ m/s}^2) = 98.1$ N of the plank and apply it at the center of mass of the plank acting downward (in the direction of gravity), or in the negative *y*-direction. (For the lowdown on self weight, head to Chapters 9 and 10.)

Writing each load in vector form

When using vector methods for solving, be sure to convert every force and moment on the F.B.D. to an appropriate vector form — Cartesian forms are usually the most common — by using the techniques I describe in Part I of this book. The following list helps you break down these forces for the example in Figure 18-1, much like the list in the preceding section does for the isolation box.

- **External forces:** To convert the moment along edge *BC,* the double-headed arrow notation (refer to Chapter 12) tells you that the direction of the moment is about the negative *z*-axis. In vector form,

$$\overline{M} = -300\hat{k} \ \text{N} \cdot \text{m}$$

  You must also convert the 100-Newton concentrated load along edge *AB* to a vector. You can use direction cosines (which I cover in Chapter 5) to determine the unit vector and then simply multiply by the 100-Newton magnitude:

$$\overline{F} = 100 \ \text{N} \cdot \left(-\sin(30)\hat{j} + \cos(30)\hat{k} \right) = -50\hat{j} + 86.6\hat{k} \ \text{N}$$

- **Internal forces:** The internal forces of the cables are unknown, but you still need to establish expressions for them in terms of their unknown magnitudes. For cable BE, you need to define a position vector from Point B (2,0,3) to Point E (0,4,3) in order to create a unit vector:

$$\overrightarrow{r_{BE}} = (0-2)\hat{i} + (4-0)\hat{j} = -2\hat{i} + 4\hat{j} \ \text{m}$$

$$\overrightarrow{T_{BE}} = T_{BE} \cdot \left(\frac{-2\hat{i} + 4\hat{j}}{\sqrt{(2)^2 + (4)^2}} \right) = -0.447(T_{BE})\hat{i} + 0.894(T_{BE})\hat{j} \ \text{N}$$

  Similarly, you can create another tension vector for cable T_{CE} using a position vector from Point C (2,0,0) to Point E (0,4,3).

$$\overrightarrow{r_{CE}} = (0-2)\hat{i} + (4-0)\hat{j} + (3-0)\hat{k} = -2\hat{i} + 4\hat{j} + 3\hat{k} \ \text{m}$$

$$\overrightarrow{T_{CE}} = T_{CE} \cdot \left(\frac{-2\hat{i} + 4\hat{j} + 3\hat{k}}{\sqrt{(2)^2 + (4)^2 + (3)^2}} \right) = -0.371(T_{CE})\hat{i} + 0.742(T_{CE})\hat{j} + 0.557(T_{CE})\hat{k} \ \text{N}$$

In both of these cable forces, even though you don't know the magnitude of the force at this time, you can still treat it as a variable and continue with your calculations. In this case, T_{BE} and T_{CE} represent the unknown magnitudes of their respective force vectors.

✔ **Support reactions:** For the roller support at Point A and the three-dimensional pinned support at Point D

$$\overrightarrow{R_A} = (A_y)\hat{j}$$

$$\overrightarrow{R_D} = (D_x)\hat{i} + (D_y)\hat{j} + (D_z)\hat{k}$$

Just as with the cable forces, you still have to include the magnitudes of the support reactions in the vectors even though they're unknown. Just leave them as variables for now — I show you how to deal with them in the following section.

✔ **Self weight:** The self weight is acting in the direction of gravity (assumed downward), so the self-weight vector is expressed as $\overline{W} = -98.1\hat{j}$ N

Writing the equilibrium equations

In Chapter 16, I introduce you to the equations that you need to develop in order to establish equilibrium conditions in three dimensions. In three-dimensional problems, you must calculate the vector resultants for forces (translational effects) and the vector resultants for moments (rotational effects), which requires a total of six equations (three each for forces and moments).

Summing forces

You establish translational equilibrium of the plank in Figure 18-1 by computing the resultant force vector of all the force vectors on the system and setting that sum equal to zero. For this example:

$$\overrightarrow{R} = \overrightarrow{F} + \overrightarrow{T_{BE}} + \overrightarrow{T_{CE}} + \overrightarrow{W} + \overrightarrow{R_A} + \overrightarrow{R_D} = 0$$

Substituting the appropriate vector equations that you computed in the previous section, and gathering all the terms with common **i, j,** and **k** directions, you can generate the equations of translational equilibrium as shown in Table 18-1, which uses units of Newton.

Table 18-1	Translational Vector Components		
	$\hat{i}$	$\hat{j}$	$\hat{k}$
F	0	−50	86.6
T$_{BE}$	$-0.447(T_{BE})$	$0.894(T_{BE})$	0
T$_{CE}$	$-0.371(T_{CE})$	$0.742(T_{CE})$	$0.557(T_{CE})$
W	0	−98.1	0
R$_A$	0	A_y	0
R$_D$	D_x	D_y	D_z

By summing each of the columns of this table, you can create the three translational equilibrium components:

i direction: $-0.447(T_{BE}) - 0.371(T_{CE}) + D_x = 0$

j direction: $-50 + 0.894(T_{BE}) + 0.742(T_{CE}) - 98.1 + A_y + D_y = 0$

k direction: $86.6 + 0.557(T_{CE}) + D_z = 0$

At this point, you have six unknown magnitudes in this problem — T_{BE}, T_{CE}, A_y, D_x, D_y, and D_z — but only three equations with which to solve them. You need to get your remaining equations from summing moments about some reference point.

Summing moments

Summing moments is probably the most labor-intensive step of solving problems with vector methods because, chances are, you'll be computing a lot of cross products (those pesky calculations I show you in Chapter 12). In fact, you need to compute one cross product for every force vector on the object. (Imagine if you had 100 forces on an object — that's a lot of matrices!) So, this stage is where you have to be extra careful in your selection of a reference point for your equivalent moment calculations.

As a general rule, select your equivalent moment reference points such that they produce the simplest mathematical expressions possible. (After all, why make your work any harder than it already is?) Places where multiple unknown forces intersect at a common point or points where dimensions are easy to compute (such as corners of objects) are often prime candidates for selection of your equivalent moment reference point. At the end of the day, the point you choose really doesn't matter from a mathematical standpoint, because your solutions produce the same numerical value — your equations just end up a lot more complex if you don't choose a convenient point.

For Figure 18-1, at Point D, you have three of your six unknowns acting simultaneously (or *concurrently*). Based on the criteria, this spot would be a good candidate for your reference because you can eliminate all three reactions at Point D from the moment equation.

Now you're ready to start computing equivalent moments. Consider each force on a case-by-case basis. You need to compute position vectors for each force on the F.B.D., so you create those by drawing your vector from the reference point (Point D) to the point of action of each force. To help you see these vectors, I've sketched each position vector for you in Figure 18-2.

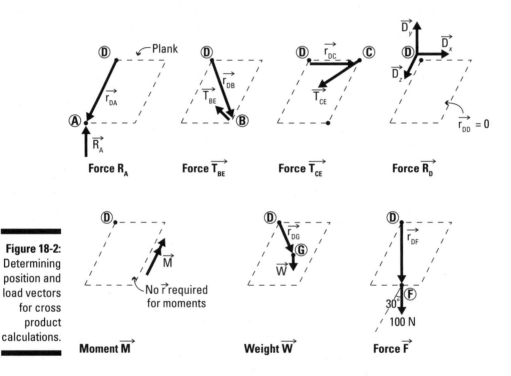

Figure 18-2:
Determining position and load vectors for cross product calculations.

Setting up to complete the cross product

The next step is to compute the equivalent moment vector equation based on the forces and moment vector of Figure 18-2. You need to perform a cross product calculation for each, and then add each of those calculations together. For this problem, the moment about D $(\mathbf{M_D})$ is written as

$$\overrightarrow{M_D} = \left(\overrightarrow{r_{DA}} \times \overrightarrow{R_A}\right) + \left(\overrightarrow{r_{DB}} \times \overrightarrow{T_{BE}}\right) + \left(\overrightarrow{r_{DC}} \times \overrightarrow{T_{CE}}\right) + \left(\overrightarrow{r_{DD}} \times \overrightarrow{R_D}\right) + \overrightarrow{M} + \left(\overrightarrow{r_{DG}} \times \overrightarrow{W}\right) + \left(\overrightarrow{r_{DF}} \times \vec{F}\right) = 0$$

You compute the position vectors for each expression just as I explain in Chapter 4.

The first letter of the subscript represents the tail point of the position vector, and the second letter of the subscript represents the head point. The moment vector **M** doesn't require a position vector because it's already a moment and it can be moved freely from one point to another on the object. Also, the term containing the position vector r_{DD} vanishes because the length of that position vector is 0. (The head and tail of your position vector are at the same point.) I selected this point for that reason — it greatly simplifies the math.

Table 18-2 shows the results of the cross product computations for each of the quantities required for M_D, the equivalent moment at D. The units for the table are Newton-meters.

Table 18-2	Calculating $\vec{r} \times \vec{F}$		
	$\hat{i}$	$\hat{j}$	$\hat{k}$
$\left(\overrightarrow{r_{DF}} \times \overrightarrow{F}\right)$	150	−86.6	−50
$\left(\overrightarrow{r_{DB}} \times \overrightarrow{T_{BE}}\right)$	$-2.682(T_{BE})$	$1.341(T_{BE})$	$1.788(T_{BE})$
$\left(\overrightarrow{r_{DC}} \times \overrightarrow{T_{CE}}\right)$	0	$-1.114(T_{CE})$	$1.484(T_{CE})$
$\left(\overrightarrow{r_{DG}} \times \overrightarrow{W}\right)$	147.15	0	−98.1
$\left(\overrightarrow{r_{DA}} \times \overrightarrow{R_A}\right)$	$-3(A_y)$	0	0
$\overrightarrow{M}$	0	0	−300

As you do with the forces, you sum each of the components in the **i, j,** and **k** Cartesian unit vector directions. From these results, you can then derive the final three equations for rotational equilibrium:

$$-3A_y - 2.682(T_{BE}) + 147.15 + 150 = 0$$

$$1.341(T_{BE}) - 1.114(T_{CE}) - 86.6 = 0$$

$$1.788(T_{BE}) + 1.484(T_{CE}) - 300 - 98.1 - 50 = 0$$

At this point, all you're left with is a mathematics problem that involves solving for the six unknowns. For this example, I would probably start with the last two expressions of the rotational equilibrium equations and solve those simultaneously for T_{BE} and T_{CE}. After you've computed those values, you can easily substitute into the remaining equations and solve for the other unknowns. The final results for this problem are

$$T_{CE} = +112.04 \text{ Newton} \qquad T_{BE} = +157.65 \text{ Newton}$$
$$A_y = -41.89 \text{ Newton} \qquad D_x = +112.03 \text{ Newton}$$
$$D_y = -34.08 \text{ Newton} \qquad D_z = -149.00 \text{ Newton}$$

As with two-dimensional problems, the positive signs on the magnitudes of this example indicate that the directions of those respective forces on the F.B.D. were in the correct directions. Negative magnitudes tell you that the forces were drawn backward.

Part VI
Statics in Action

The 5th Wave   By Rich Tennant

"I love it when Professor Buttonhole teaches static equilibrium."

In this part . . .

This part gets down to showing you the different types of structures you encounter in your statics work. Each chapter focuses on the different solution techniques that are unique to a given system type, including breaking up a problem and sketching the various free-body diagrams you need in order to write the equations of equilibrium.

Chapter 19

Working with Trusses

Many common structural systems that are used in buildings are made from numerous members that have been connected together to form a more complex system. When you enter your local hardware store or warehouse shopping center, take a look up at the ceiling. You may see the most popular type of these structural systems: trusses.

Trusses are very lightweight structural systems, capable of spanning very long distances. They're used to span major rivers or to span football fields or basketball courts in arenas. You may even possibly have trusses in the roof of your home. Trusses provide a wide array of shapes and sizes that make them extremely versatile to engineers and architects.

I start this chapter by identifying the major criteria that define trusses so that you can spot a truss when you see one. Then I introduce you to two of the most popular methods of solving for internal forces in the members of the trusses: the method of joints and the method of sections. I conclude the chapter by showing you how to determine zero-force members in trusses without ever writing a single equation.

Identifying Truss Members

Trusses are structural systems that are composed of numerous members connected together. You may encounter a wide variety of shapes in a truss, but these shapes always have several common basic properties:

✔ **Internal forces:** All members in a truss system must be axial-only members — that is, their internal forces are *axial* (parallel to each member's longitudinal axis) and defined in terms of tension and compression.

✔ **End connections:** All members in a truss are pinned at the ends at locations known as *joints* (sometimes referred to as panel points). It's these pinned ends that ensure all internal forces in truss members remain axial.

✔ **Load locations:** All loads are concentrated loads (see Chapter 9) and can only be applied at the joint locations. Truss systems have no distributed loads or concentrated moments. If a concentrated load isn't applied at the joint, you can't analyze the structure as a truss; instead, you need to analyze it as a frame or machine, which I introduce in Chapter 21.

Trusses are especially useful because of the variety of ways that the members can be put together. Figure 19-1 shows examples of several trusses. Notice that straight members can be used to create a wide variety of roof shapes. Even curved and peaked roofs can be created with trusses. If you can imagine it, you can probably create it with a truss system.

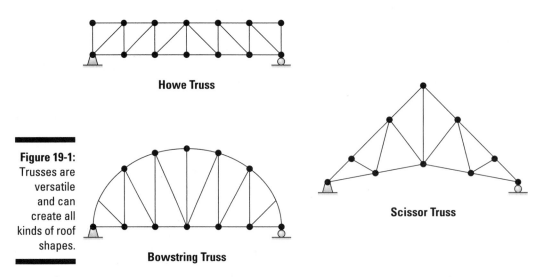

Howe Truss

Figure 19-1:
Trusses are versatile and can create all kinds of roof shapes.

Bowstring Truss

Scissor Truss

When you're solving a truss problem, the first steps are no different than the first steps of any other statics problem. You always start a statics problem by drawing a free-body diagram (F.B.D. — check out Part IV) of the entire system; you then write the equilibrium equations to determine as many of the support reactions as possible.

In this chapter, I provide all the support reactions and skip drawing the initial F.B.D. in order to help you focus on the different solution techniques for trusses. Just don't forget to first find the magnitudes of the support reactions when you work your own problems.

When you have the support reactions calculated, you're ready to start solving for internal forces by using one of the two common methods: the method of joints or the method of sections. Each of these methods has benefits and drawbacks. I explain both in the following sections.

The Method of Joints: Zooming In on One Panel Point at a Time

The first method for solving a truss system is pretty straightforward. As the name of the method implies, you draw an F.B.D. for each joint in the truss; you then apply the equilibrium equations to those diagrams.

Consider the truss shown in Figure 19-2, which is pin-supported at Joint A and roller-supported at Joint E. The support reactions have been given. With this figure, I show you how to use the method of joints to solve for the internal truss forces in the following sections.

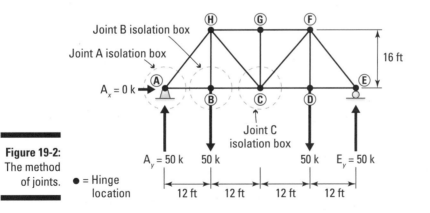

Figure 19-2: The method of joints. ● = Hinge location

Step 1: Drawing isolation boxes

To start the analysis of this truss by the method of joints, first draw F.B.D.s of each joint by drawing an isolation box around each joint. Draw an isolation box to isolate the joint of interest. When you isolate a joint, you also end up cutting each truss member that's connected to that joint. Also include any support reactions or applied loads that show up on the F.B.D. at that location. Figure 19-3 shows the F.B.D. for Joints A, B, and C.

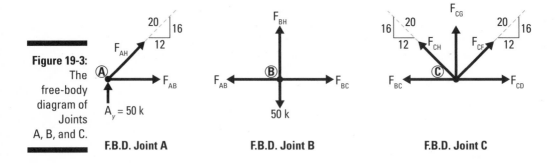

Figure 19-3:
The
free-body
diagram of
Joints
A, B, and C.

F.B.D. Joint A   F.B.D. Joint B   F.B.D. Joint C

When drawing the internal forces for trusses, I find it convenient to assume that unknown internal member forces are always in tension. At the joint locations, by assuming a member is in tension, you must draw the internal force as though it's pulling on the joint.

The following list explains the method you use to draw the F.B.D.s for each of the joints at A, B, and C.

- ✔ **F.B.D. for Joint A:** At Joint A, the isolation box cuts members AB and AH. To account for these revealed internal forces, you draw F_{AB} with the force arrow point to the right, and you draw F_{AH} with the arrow point up and to the right on the joint of A in the F.B.D. You also have the support reaction A_y = 50 kip acting upward at Joint A. The horizontal reaction A_x = 0 kip equals zero, so I've omitted it from the F.B.D.

- ✔ **F.B.D. for Joint B:** At Joint B, the isolation box cuts three members — AB, BH, and BC — so you must also include the internal forces for each of these three members on your F.B.D. For Joint B, the internal force F_{AB} must now be pointing to the left in order to be pulling on Joint B. There is also an applied load of 50 kip acting downward at this joint, so you include that on the F.B.D. of the joint as well.

- ✔ **F.B.D. for Joint C:** At Joint C, the isolation box cuts a total of five members — BC, CH, CG, CF, and CD — revealing a whopping total of five internal forces on the same joint. Because you have only three equilibrium equations in two-dimensions, this joint is two degrees indeterminate, which means you need to find several of these forces by other means (from other joint F.B.D.s).

On each of these free-body diagrams, you also want to make sure to include the angles or proportion triangles for all of the forces. (Head to Chapter 5 for more on proportion triangles.) However, if the force is horizontal or vertical, you don't really have to indicate the direction because you already know that the line of action is oriented at 0 degrees for a horizontal force and at 90 degrees for a vertical force.

Now, if you're observant, you'll notice that $\mathbf{F_{AB}}$ has shown up on two free-body diagrams. On Joint A, it's pointing to the right, and in Joint B, it's pointing to the left. How can the same internal tension force be pointing in two different directions? Look at an F.B.D. of member AB and its connecting joints, shown in Figure 19-4.

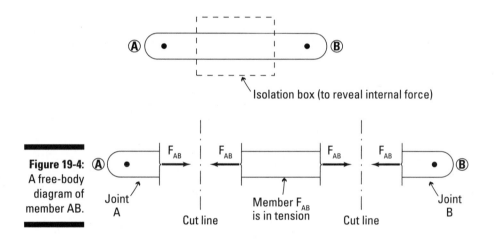

Figure 19-4:
A free-body diagram of member AB.

To reveal the tension and compression forces in member AB, you start by drawing an isolation box (which I cover in Chapter 14) that cuts the member at two locations as shown. If you assume that the member is subjected to tension (as I do), in order for this member to be in tension, the force $\mathbf{F_{AB}}$ must be drawn as shown — with the left end of the cut member pointing to the left, and the right end of the cut member pointing to the right.

Remember that internal forces must remain in equilibrium on either side of a cut line. That means that the part of the member that's connected at the joint has an internal force that must balance the $\mathbf{F_{AB}}$ force on the other side of the cut line from it. For Joint A, the internal force is balancing the force on the left end of the cut member that's pointing to the left. So, on Joint A, it must be pointing to the right, or pulling on the joint as you assumed. Similarly, for Joint B, the internal force at this joint must balance the force $\mathbf{F_{AB}}$ that is acting on the right end of the cut member, which is pointing to the right. That means that the internal force of member AB acting on Joint B is pointing to the left, or pulling on the joint.

TIP

This logic is applied at every joint a member is connected to. For each of the two joints that connect a member, the assumed direction of forces is always in opposite directions — even on diagonals.

Step 2: Applying the equations of equilibrium

After you've drawn the F.B.D.s for a particular joint, your next step is to write the equations of equilibrium, which I introduce in Chapter 16.

$$+\uparrow \sum \left|\vec{F_y}\right| = 0 \rightarrow A_y + \frac{16}{20}F_{AH} = 0 \rightarrow \frac{16}{20}F_{AH} = -50 \text{ k}$$
$$\Rightarrow F_{AH} = -62.5 \text{ k} \rightarrow F_{AH} = 62.5 \text{ k}(C)$$

From this equation, you can solve for the internal force $\mathbf{F_{AH}}$, which turns out to have a negative value. As in all equilibrium problems, this fact means that the assumed direction of the internal force on the F.B.D. was incorrect. Because you assumed it was in tension, the negative sign means that it's actually in compression — that's why the final answer has a (C) after it. This way, on all F.B.D.s that contain the force $\mathbf{F_{AH}}$, you know that force is in compression. If you find that a member is in tension, you write a (T) after it.

Next, you can write the other equilibrium equation for Joint A.

$$\pm \sum \left|\vec{F_x}\right| = 0 \rightarrow F_{AB} + \frac{12}{20}F_{AH} = 0 \rightarrow F_{AB} = -\frac{12}{20}\left(F_{AH}\right) = 0$$
$$\Rightarrow F_{AB} = -\frac{12}{20}\left(-62.5 \text{ k}\right) \rightarrow F_{AB} = +37.5 \text{ k} \rightarrow F_{AB} = 37.5 \text{ k}\left(T\right)$$

Notice that this equation was written for the exact way that I originally drew the F.B.D. at Joint A. This equation is based on the assumption that the force in member AH, $\mathbf{F_{AH}}$, is still in tension, even though you now know that's not the case. But that's okay — trust me.

All you have to do now is simply plug the signed valued into the equation. If you calculated a positive value, you plug in a positive value, but because $\mathbf{F_{AH}}$ was negative, you actually substitute the negative value into the equation. The signed values take care of any errors with your assumed force direction within the equilibrium equation.

Finally, you can also write the moment equation for Joint A:

$$\zeta + \sum \left|\vec{M_A}\right| = 0 \rightarrow F_{AH}\left(0\right) + F_{AB}\left(0\right) + A_y\left(0\right) = 0 \rightarrow 0 = 0$$

As you can see, each force on this F.B.D. is acting *concurrently* (simultane-ously) through Joint A. By summing moments at this point, the perpendicular distance to each force is actually zero. So for the method of joints, the moment equation no longer gives you any useful information about relation-ships between forces.

The loss of the moment equation's usefulness isn't a major hindrance. It just means that in order for you to work with a particular joint, you can only completely solve all the forces if there are no more than two unknown forces acting on the F.B.D. of interest.

Step 3: Review and repeat

Your next step is to look for another joint that has only two unknowns. At this time, the F.B.D. of Joint B has two unknown forces (just F_{BH} and F_{BC}, because the preceding section determined F_{AB}), so you can move to that joint and repeat the process.

You keep moving from one joint to the next, learning more about the different member forces, and applying them to each F.B.D. Repeating this for all joints in the truss reveals all the internal forces within each member.

Drawbacks to the Method of Joints

The major drawback with the method of joints is that in order to solve for a member force that's connected to a joint with a large number of other mem-bers, such as Joint C in Figure 19-3 earlier in the chapter, you have to first solve for forces at several other places. In fact, to solve for the force F_{CD}, you have two options:

> ✔ **Starting at Joint A:** First, solve for forces at Joint A as described in the preceding section. Next, go to Joint B and do the same. Then go to Joint H, followed by Joint G, at which time you could move to Joint C and finally solve for force F_{CD}.

> ✔ **Starting at Joint E:** First, solve for forces at Joint E, followed by Joint D.

Depending on which joint you start with, you can greatly increase the number of free-body diagrams you have to work with. In this case, to find F_{CD} by starting at Joint A, you have a minimum of five joints, or ten translational equilibrium equations you have to write. Starting at Joint E has much fewer F.B.D. stops along the way, but you still have to work at two places, or four total translational equilibrium equations.

What happens if the truss you're analyzing is very long and has hundreds of joints? If you're interested in a force in the middle of this truss, you may end up having to write an extremely large number of equations. And if you have irrational numerical values for each force along the way, you may end up incorporating a lot of error into your final answer. I won't even mention the fact that the more steps you take, the more chances you have for writing an incorrect equation or keying a bad numerical value into your calculator. Generally speaking, the fewer calculations you have to do to get a desired answer, the more likely that you won't make a mistake along the way.

To find forces in the middle regions of long trusses, a technique that lets you skip directly to the middle would be pretty handy. That's where the second major method of truss analysis, the method of sections, comes into play (see the following, well, section).

And Now for My Next Trick: Slicing through the Method of Sections

The second major method of truss analysis is the method of sections. As the name implies, in this method you basically slice a truss into sections or, more specifically, into two pieces. But you can't just go hacking up the truss — you need to obey several rules:

- **Cut the member that you're interested in.** Obviously, to calculate the internal force of a truss member, you must first expose the internal force. The only way you can do that is by cutting the member.

- **Cut the truss completely into two parts.** Don't cut partway through a truss and then stop. In fact, to assure yourself, you may consider using an isolation box that's completely closed. *Remember:* You can include one of the supports inside your box.

- **Cut a maximum of three members.** You're allowed to cut up to three members total. If you cut fewer than three, that's fine. But if you cut four or more, you may have some problems with this method because you may not be able to solve for all of the unknowns on the free-body diagram.

- **If you cut more than three members, all but two of the members must be on the same line of action.** In rare circumstances, you can actually cut more than three members, but you'll need to make sure that all but at least two of them are collinear. In general, it's a very rare occasion when you can actually cut more than three members at a time.

If performed properly, the method of sections can save you a load of time analyzing those members in the middle regions of trusses. In fact, in many cases you can actually calculate the force in a particular member by writing just a single equation.

As with any truss analysis, the first steps require that you draw an F.B.D. and compute as many of the external support reactions as possible. However, with the method of sections, if you have a truss that's *cantilevered* (or supported in such a way that at least one end of the truss is unsupported), you may not even have to find the support reactions at all — although calculating support reactions is always a good idea, if you can.

Step 1: Cutting the truss

Suppose you're tasked with the job of finding the force in member CD of the truss shown in Figure 19-5.

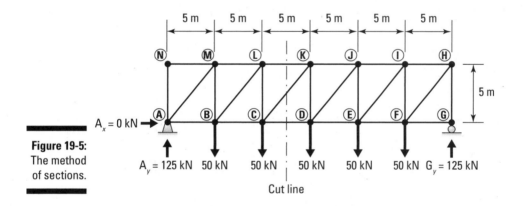

Figure 19-5: The method of sections.

If you use the method of joints (see "The Method of Joints: Zooming In on One Panel Point at a Time" earlier in the chapter) to compute this force, you need no fewer than six free-body diagrams (start at N, A, M, B, L, and finally C) or even seven free-body diagrams (start at G, H, F, I, E, J, D), depending on which end of the truss you started on. That's a lot of diagrams and even more equation writing. (***Remember:*** You have to write two equations per F.B.D. for the method of joints.)

To analyze the truss of Figure 19-5, a likely way to cut this truss would be to cut members LK, CK, and CD (which is your desired member). This strategy obeys all the method-of-sections rules I explain earlier in the chapter.

Step 2: Drawing the F.B.D. for the two remaining truss pieces

After you cut the truss, you're left with two pieces. Draw the free-body diagrams, including all reactions, point loads, and the revealed internal forces of members LK, CK, and CD, as shown in Figure 19-6.

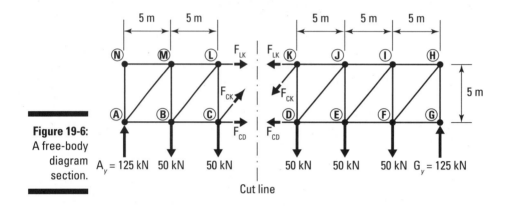

Figure 19-6: A free-body diagram section.

Step 3: Applying the equations of translational equilibrium

The F.B.D. that you choose to work with doesn't really matter, but I usually recommend choosing the smaller of the two remaining pieces because this strategy usually ensures that your equilibrium equations are smaller as well. For this example, I work with the left F.B.D. in Figure 19-6 because it has fewer applied loads. However, if you choose to work with the F.B.D. on the right instead, you still end up calculating the exact same internal forces, so which diagram you select doesn't really matter.

In this step, you write the two translational equilibrium equations.

$$\xrightarrow{+}\sum\left|\vec{F_x}\right| = 0 \rightarrow F_{LK} + \left(\frac{5\text{ m}}{7.07\text{ m}}\right)\cdot F_{CK} + F_{CD} = 0$$

$$+\uparrow\sum\left|\vec{F_y}\right| = 0 \rightarrow 125\text{ kN} - 50\text{ kN} - 50\text{ kN} + \left(\frac{5\text{ m}}{7.07\text{ m}}\right)\cdot F_{CK} = 0 \rightarrow F_{CK} = -35.4\text{ kN}$$

$$\Rightarrow F_{CK} = 35.4\text{ kN}\left(C\right)$$

The translational equilibrium equations give you an internal force for a member in the interior part of the truss, just not the one you're interested in at this time. But never fear — you still have one equilibrium equation left to work with.

Step 4: Applying the equation of rotational equilibrium

The last equation you have to work with is the moment equation. The moment equation is typically the most useful because you have complete control over the point about which the equation is written. In the method of sections, the moment equation truly shines.

To choose the point for the moment equation, you need to look back at the F.B.D. In this system, you have three unknown internal forces acting on the truss after you cut it, one of which is the force you're interested in. So it would be convenient to write an expression that includes $\mathbf{F}_{CD}$ but doesn't include $\mathbf{F}_{LK}$ or $\mathbf{F}_{CK}$ in it. (You know that if you calculate the moment of a force at a point on its line of action, the moment from this force is actually zero.) Thus the only way to make that happen is to write the moment equation somewhere along the lines of action of both member LK and member CK.

Because these two members have different lines of action, only one point in space is common to both members; the intersection of the lines of action of member LK and CK happens to be at Joint K. Both of those two forces intersect at that location. So if you sum moments at that location, both forces will have a perpendicular distance of 0 meters because that point is concurrent with each of the forces. Summing moments at Joint K, you're now left with

$$\circlearrowleft + \sum \left| \overrightarrow{M_K} \right| = 0 \rightarrow -125 \text{ kN}(15 \text{ m}) + 50 \text{ kN}(10 \text{ m}) + 50 \text{ kN}(5 \text{ m}) + F_{CD}(5 \text{ m}) = 0$$
$$\Rightarrow F_{CD} = +225 \text{ kN} \rightarrow F_{CD} = 225 \text{ kN}\,(\text{T})$$

Check it out! You've found the force of a member in the middle of the truss by writing only one equation. All this was possible because you chose to calculate your moment at a convenient point where the other forces intersect, and you stuck to the basic rules for creating the F.B.D.

Choosing a convenient point about which to write your equations isn't just useful in statics but is also a very handy concept in many other areas of study, such as structural analysis, dynamics, and mechanics of materials, among others. So keep your eyes open for those special locations.

Step 4, continued: Identifying the instantaneous center

Instantaneous center is a term used to describe those locations on an object where special things happen. In dynamics, the term refers to the point in space (or on an object) about which all other points are moving, resulting in a velocity of zero. Although this terminology isn't 100 percent correct for statics, the term can still be used to describe the point through which multiple forces acting result in zero moment.

For the method of sections, the instantaneous center is the point you look for in order to ensure the internal force you're looking for appears, while making the other unknown forces disappear from your moment equation. In many cases, especially with trusses of horizontal chords, the instantaneous center is often at one of the joints of the truss. However, in practice, this point doesn't have to be at a joint. In fact, it doesn't have to be within the truss at all. Consider the simply supported truss, known as a Gambrel truss, shown in Figure 19-7. (For more on Gambrel trusses, see the nearby sidebar.)

Figure 19-7:
An example
of instan-
taneous
center.

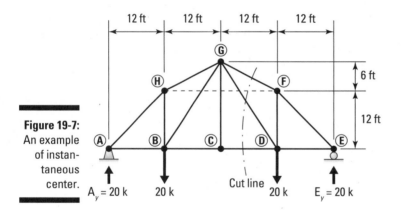

Suppose you wanted to determine the force in truss member DG of the gambrel truss shown in Figure 19-7. Because this member is in the middle of the truss, your first instinct should be to look at the method of sections as a possible solution and to slice the truss on the cut line through member DG. One possibility is to cut through members FG, DG, and CD.

Taking the F.B.D. on the right half, shown in Figure 19-8, you can see that the lines of action for members GF and CD actually intersect at a point outside the boundary of the truss. I define that "extra" distance beyond Joint E as *x*.

A brief history of Gambrel trusses

The truss shown in Figure 19-7 is known as a Gambrel truss, which is known for having a *double-pitched* top chord, or a top chord that has more than one slope associated with it.

The Gambrel roof is thought to have been developed in Indonesia before it was brought to the Americas by French, Spanish, English, and Dutch explorers in the early- to mid-1700s. The shape served a function in that it also allowed for easy access of supplies into the upper areas, or allowed smoke to escape from cooking ovens located inside the structure.

The shape of this truss has a very classic profile to many of agricultural structures of the American Midwest from the 1800s. If you ask someone to draw a picture of a typical barn, chances are he'll draw a Gambrel roof. In many countries around the world, Gambrel trusses (sometimes referred to as Mansard trusses, despite a subtle difference in actual construction) are still commonly used today.

To find the distance x, you need to construct a set of similar triangles that incorporate the slopes of the two members you want to eliminate. In this case, you want to know the force in member DG, F_{DG}, which means that you want to eliminate forces F_{FG} and F_{DC} from your moment equation. These are the two forces that you need to work with to find the instantaneous center at Point O. To accomplish this, you set up a relationship using similar triangles.

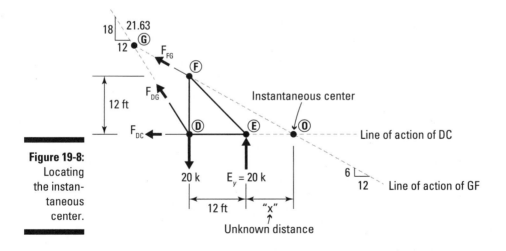

Figure 19-8: Locating the instantaneous center.

For member force $\mathbf{F_{FG}}$, you know that its line of action has a 6:12 proportion (height-to-length ratio) to describe the slope. This line of action is also the hypotenuse of $\triangle ODF$. Similarly, member force $\mathbf{F_{DC}}$ is horizontal and forms the lower edge of the same triangle. $\triangle ODF$ has a height of 12 feet (side DF) and a length of $(x + 12)$ feet for side DO, or a ratio of $12:(x + 12)$. Knowing that these ratios must be the same, you can then set up a relationship and solve for x:

$$\frac{6 \text{ ft}}{12 \text{ ft}} = \frac{12 \text{ ft}}{x + 12\text{ft}} \Rightarrow x = 12 \text{ ft}$$

Thus, the instantaneous center at Point O is located at a distance of 12 feet to the right of Point E. When you know this location, you can write the sum of moments equation at Point O and solve directly for the force in member F_{DG}:

$$\circlearrowleft + \sum \left|\overline{M_o}\right| = 0 \rightarrow 20 \text{ k}(24 \text{ ft}) - 20 \text{ k}(12 \text{ ft}) - \left(\frac{18}{21.63}\right) \cdot F_{DG}(24 \text{ ft}) = 0$$

$$\Rightarrow F_{DG} = +12.0 \text{ k} \Rightarrow F_{DG} = 12.0 \text{ k} \, (\text{T})$$

Shortcutting the Equation Writing: Zero-Force Members

For some very specific cases, you can determine the forces within a truss without ever performing a single calculation. You just need to be able to identify a couple of conditions, and if all those conditions are met, you can draw conclusions about whether a force is considered a zero-force member, without a doubt. To identify a zero-force member, you have to look at each joint on a case-by-case basis and count the number of members, external reactions, or applied point loads that are acting at the current joint being investigated. Figure 19-9 and the following list illustrate the three specific cases that allow you conclude whether a member is zero force.

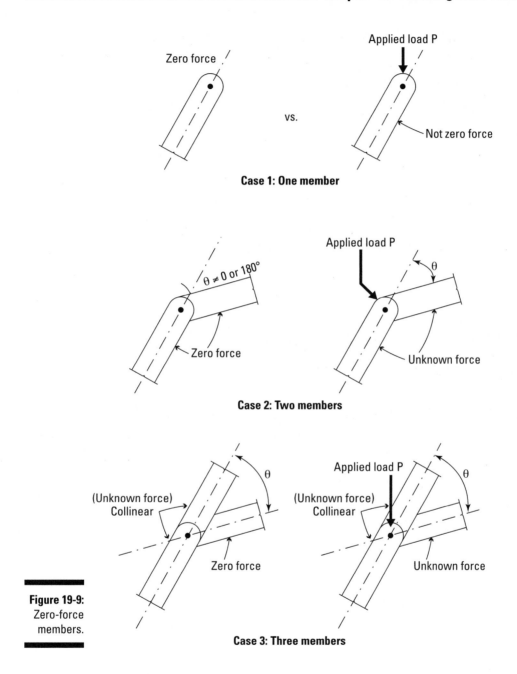

Figure 19-9:
Zero-force
members.

- ✔ **One member:** If a joint has only one member and no point loads acting at the joint, the member must be zero force.

- ✔ **Two members:** If a joint has two members acting at it, and no point load is applied at the joint, both are zero force as long as the members aren't collinear. If the angle between the members is either 0 degrees or 180 degrees, you don't know for certain that both are zero, so you have to leave them in your computations.

- ✔ **Three members:** If a joint has three members acting at it, two of which are collinear (and one of which isn't), and no point load is applied, the noncollinear member is zero force. You don't know the internal forces for the two collinear members for sure, so you must retain them in your calculations.

If you have a joint that has two members and a point load, a special case may exist. If the point load is collinear with a member, the noncollinear member is zero force. This case is useful for members at support locations. Often, the members are perpendicular or parallel to support reactions.

When you've concluded that a member is a zero-force member, remove it from your F.B.D. of the entire system, and then go back and look at the system again. Sometimes, removing one member from the F.B.D. helps you conclude that another member is also zero by these same basic definitions.

Also, you only remove a member from the analysis if you're absolutely sure it's zero force. If the joint doesn't meet the requirements in this section, or you're still unsure, don't remove it. Your calculations in the end will verify the member is zero force anyway.

In fact, you usually want to identify whether a member is zero force before you even start employing the method of joints or the method of sections. You typically start looking for zero-force members after you have calculated the support reactions. Being able to identify zero-force members can be especially useful because you can remove members from the analysis altogether, which can greatly simplify your free-body diagrams.

Chapter 20

Analyzing Beams and Bending Members

*F*or many engineers, the ability to analyze the internal forces of bending members is probably one of the most important skills that you can develop, partly because many structures and objects in the real world aren't just axially loaded. In fact, the beams in the floor you're sitting on (assuming it's not the ground floor) or even the rafters in your roof are all examples of bending members.

In this chapter, I help define what a bending member is, and I show you what to look for when you identify it. When you know you're dealing with a bending member, you're ready to actually compute the magnitudes of internal bending forces, and I show you how. I also decipher yet another sign convention (assumption) and show you the steps to develop generalized equations for internal loads. Finally, I show you one of the most important tools for engineers: the ability to quickly draw diagrams of these internal forces.

Defining the Internal Bending Forces

In Chapter 19, I show you how to handle truss systems, which are comprised of members subjected to only axial internal forces, which cause a member to become longer in tension and shorter in compression (see Figure 20-1).

But what happens when a member isn't loaded with just an axial load? Figure 20-1 also shows a point load that is acting perpendicular to the member's longitudinal axis. When this type of member (also known as a *bending member*) is loaded, it wants to deflect in the direction of the applied load, but not in the direction of the longitudinal axis of the member. As a result, axial tension or compression are no longer the only forces that appear internally.

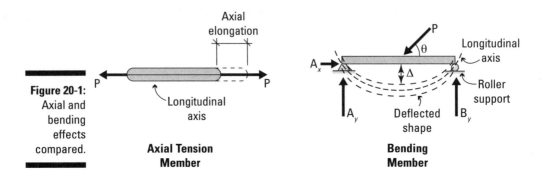

Figure 20-1:
Axial and
bending
effects
compared.

Axial Tension Member

Bending Member

And then there were three: Internal forces of two-dimensional objects

To start the investigation, you need to cut the bending member to expose the juicy goodness inside, as I show in Figure 20-2 at an arbitrary distance x from the left support. When you cut a bending member, instead of just an axial force, three internal loads will appear to help maintain equilibrium.

- ✔ **Axial:** By inspecting Figure 20-2a, to balance the horizontal reaction A_x, there must be an internal axial force N_x that acts parallel to the longitudinal axis to ensure translational equilibrium in this direction.

- ✔ **Shear:** By inspecting Figure 20-2b, you can see that without the presence of the force V_x, the free-body diagram (F.B.D. — see Part IV) of the object can't be in translational equilibrium in that direction. This internal force V_x is also known as a shear force, and it acts perpendicular to the longitudinal axis of the member.

- ✔ **Moment:** From Figure 20-2c, you can see that without the presence of the moment M_x the F.B.D. can't be in rotational equilibrium about Point X. The effect of the vertical support reaction and the shear force actually causes a rotational behavior that must be balanced by the internal moment. The contribution of the axial force N_x to the internal moment is often neglected because the perpendicular distance, Δ, is generally very, very small.

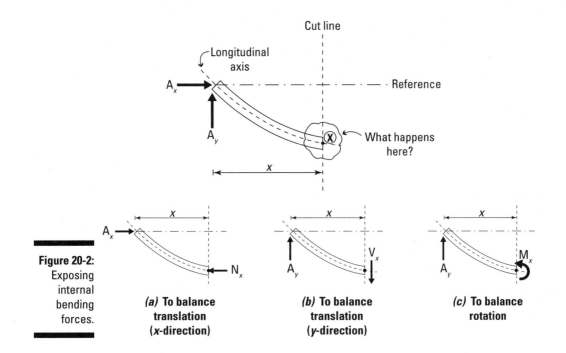

Figure 20-2:
Exposing
internal
bending
forces.

(a) To balance
translation
(*x*-direction)

(b) To balance
translation
(*y*-direction)

(c) To balance
rotation

Strange new three-dimensional effects

As I show in Figure 20-3, in three dimensions, six internal loads appear on
every exposed surface (or section) — one for each equation of equilibrium
(flip to Chapter 16).

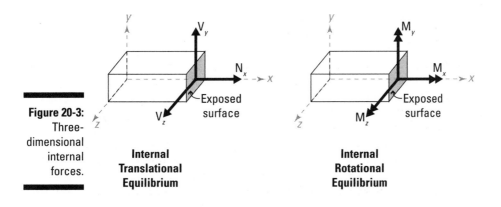

Figure 20-3:
Three-
dimensional
internal
forces.

**Internal
Translational
Equilibrium**

**Internal
Rotational
Equilibrium**

Translation: Another shear force

To balance the exposed internal loads in a three-dimensional F.B.D., you need three noncoplanar translational forces to provide the necessary forces for equilibrium. You always have an axial force N_x that acts along the longitudinal axis of the member, and you also have a vertical shear force, labeled V_y. The third internal force V_z also acts perpendicular to the longitudinal axis — which makes it an additional shear force (refer to Figure 20-3).

Rotation: Torsion and another bending moment

There must also be three rotational equilibrium requirements in a three-dimensional free-body diagram (refer to Figure 20-3). Just as with two-dimensional problems, a moment about the z-axis M_z is necessary. There must also be a *torsional moment* (sometimes referred to as a twisting moment) about the longitudinal axis M_x. The third moment that is required is an additional bending moment, M_y.

Calculating Internal Loads at a Point

When beams are designed to be used in buildings, these internal loads are what engineers must design for. Engineers must follow local building codes and guidelines and have a thorough understanding of the materials that they are working with. At this point, I just stick with showing you how to crunch the numbers.

To calculate internal loads at a point, you simply cut your structure at the point of interest, draw the F.B.D. (including the newly exposed internal forces), and then apply the equations of equilibrium (sound familiar?). However, with internal loads, you want to be very mindful of the sign convention that you select.

Positive moments make you happy!: Yet another two-dimensional sign convention

When you expose an axial force on an axial-only member, the sign of the force depends on which side of the cut (or exposed face) you're working with. An axial force that is positive in one direction for one F.B.D. would be negative for the F.B.D. on the other side of the cut line (or exposed surface). Figure 20-4 shows the sign convention for two-dimensional bending members.

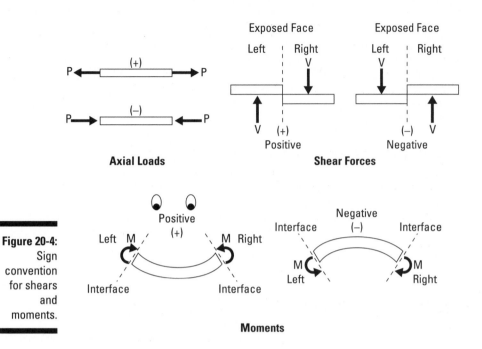

Figure 20-4:
Sign convention for shears and moments.

✔ **Axial forces:** The sign of the internal axial forces depends on whether the object of the F.B.D. is subject to a tensile or compressive load. The sign convention for axial forces from bending effects is the same. Tension is positive and compression is negative.

✔ **Shear forces:** If the force is acting on the left side of an interface or exposed face, it's positive if it acts upward. If the force is acting on the right side of the interface or exposed face, it's positive if it acts downward.

✔ **Bending moments:** A positive bending moment on each end of a section causes a beam to flex downward in the middle. A clockwise moment on the left side of the interface (or cut line) is considered positive, while a counterclockwise moment on the right side of the cut line is also considered positive.

Now, if you picture the positive moment deflected shape as the mouth, and the (+) indicator as a nose, all you need are a couple of eyes on your sketch, and you'll have created a happy face. That's why you can say, "Positive moments make you happy!"

Using the sign convention

Often, you don't know the direction of an internal load until after you perform your equilibrium calculations. Just make sure to include the internal loads on your F.B.D.s at the beginning, and let your equilibrium calculations confirm your assumed directions. It usually helps to be consistent, so normally you want to assign the directions of your internal forces based on the positive sign convention that I discuss in the preceding section.

Consider the basic portal frame of Figure 20-5 which experiences a 50-kilo-Newton (kN) force, a 2 kN per-meter distributed force, and a 40 kN per-meter concentrated moment. Suppose you want to compute the internal forces acting at Point E of the beam BC. In this example, the support reactions and directions at Points A and D have already been shown and are included on the drawing. *Self weight* (gravitational effects on the system) isn't a concern for this problem.

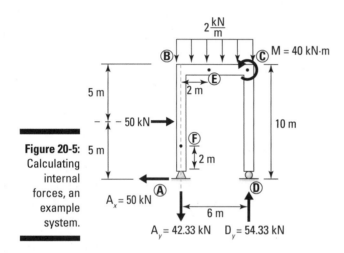

Figure 20-5: Calculating internal forces, an example system.

To calculate internal forces, the first step is to cut the beam at the location of interest (Point E, for this example), and draw a free-body diagram of one of the pieces that remains. F.B.D. #1 in Figure 20-6 shows the left half of the structure with all the applied external loads, support reactions, and appropriate dimensions. Remember, you also need to include the internal forces — an axial force, a shear force, and an internal moment.

The assumed direction of the internal loads is determined by looking at the cut line. From the F.B.D. #1 of Figure 20-6, because all of the internal loads are acting to the right of the cut line, you look at the "right" side of the sign convention diagram in Figure 20-4 for each of the internal forces.

The axial force is assumed to be positive if it's pulling (or acting tensile) on the longitudinal axis of the cut member. In this case, N_E represents the axial force on F.B.D. #1 and is acting to the right. In order for the internal shear force V_E (which is acting on the right side of the cut line) to be positive, it must be acting downward (which is perpendicular to the longitudinal axis). The bending moment M_E (also on the right side of the exposed face) must be acting in a counterclockwise direction to be positive.

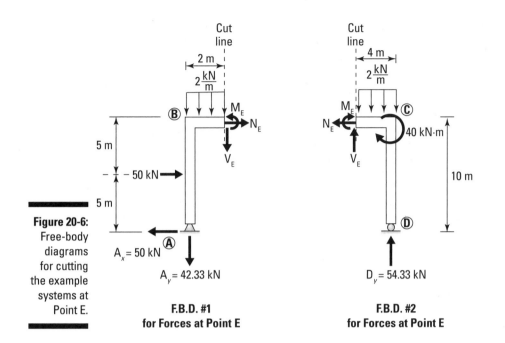

Figure 20-6:
Free-body diagrams for cutting the example systems at Point E.

F.B.D. #1
for Forces at Point E

F.B.D. #2
for Forces at Point E

For reference, F.B.D. #2 in Figure 20-6 shows the other half of the structure, with the appropriate loads, moments, support reactions, and dimensions already included on that F.B.D. For this diagram, the same internal loads have been revealed, but this time they're acting on the left side of the cut line which means they're applied with their senses reversed from those of F.B.D. #1. On F.B.D. #2 of Figure 20-6, the axial load N_E, is in tension when it acts to the left, the internal shear V_E is assumed positive when it acts upward, and the bending moment M_E is now acting positively in a clockwise direction.

Notice that the internal loads drawn on a F.B.D. on one side of a cut line are always equal and opposite to the internal loads drawn on the F.B.D. on the other side of the cut line. This ensures that the equilibrium is maintained at the cut line.

Computing internal force magnitudes

With the internal forces drawn, you're now ready to employ the equations of equilibrium that I show you in Part V. For now, I use F.B.D. #1 of Figure 20-6 for my calculations.

REMEMBER

It really doesn't matter which of the two free-body diagrams of Figure 20-6 you choose to work with. Assuming that you've calculated the support reactions correctly and that you have all the applied loads on both free-body diagrams and at their proper locations, the numerical calculations you perform produce the same results.

I start by summing forces in the x-direction (assuming to the right as positive).

$$\xrightarrow{+} \sum \left| \overrightarrow{F_x} \right| = 0 \rightarrow +N_E + 50 \text{ kN} - A_x = 0 \rightarrow +N_E + 50 \text{ kN} - 50 \text{ kN} = 0$$
$$\Rightarrow N_E = 0 \text{ kN}$$

At Point E, the axial load, $N_E = 0$ kN. Don't worry that the axial value is computed to be zero. Numerically, internal forces are permitted to be zero. Next, you compute the internal shear force at Point E:

$$+\uparrow \sum \left| \overrightarrow{F_y} \right| = 0 \rightarrow -A_y - \left(2 \text{ kN/m} \right) \cdot \left(2m \right) - V_E = 0 \rightarrow -42.33 \text{ kN} - 4 \text{ kN} - V_E = 0$$
$$\Rightarrow V_E = -46.33 \text{ kN}$$

The negative value on the magnitude of the shear V_E indicates that the direction that was assumed on the free-body diagram. Finally, you compute the internal moment at Point E.

$$\curvearrowleft + \sum \overline{M} = 0 \rightarrow +M_E + 50 \text{ kN} \cdot \left(5 \text{ m} \right) - A_x : \left(10 \text{ m} \right) + ..$$
$$+ A_y \cdot \left(2 \text{ m} \right) + \left(2 \text{ kN/m} \right) \cdot \left(2 \text{ m} \right) \cdot \left(\frac{2 \text{ m}}{2} \right) = 0$$
$$\Rightarrow +M_E + 250 \text{ kN} \cdot \text{ m} - 50 \text{ kN} \cdot \left(10 \text{ m} \right) + 42.33 \text{ kN} \cdot \left(2 \text{ m} \right) + 4 \text{ kN} \cdot \text{ m} = 0$$
$$\Rightarrow M_E = +161.33 \text{ kN} \cdot \text{ m}$$

The magnitude of the internal moment, M_E, is computed to be +161.33 kN-meters. Because this value is positive, the assumed direction is correct.

I can also compute the magnitude of the internal moment, M_E, using F.B.D. #2:

$$\curvearrowleft + \sum \overline{M} = 0 \rightarrow -M_E + D_y \cdot \left(4 \text{ m} \right) - \left(2 \text{ kN/m} \right) \cdot \left(4 \text{ m} \right) \cdot \frac{4 \text{ m}}{2} - 40 \text{ kN} \cdot \text{ m} = 0$$
$$\Rightarrow -M_E + \left(54.33 \text{ kN} \right) \cdot \left(4 \text{ m} \right) - \left(2 \text{ kN/m} \right) \cdot \left(4 \text{ m} \right) \cdot \left(\frac{4 \text{ m}}{2} \right) - 40 \text{ kN} \cdot \text{ m} = 0$$
$$\Rightarrow M_E = +161.33 \text{ kN} \cdot \text{ m}$$

Thus, the internal bending moment calculations produce the same result, regardless of which F.B.D. you decide to work with. The only difference is that the directions (or the senses) of the forces are equal and opposite.

Writing Generalized Equations for Internal Forces

If you want to calculate the magnitudes of internal forces at specific points, the procedure I outline in the preceding section is more than adequate. Unfortunately, if you want to properly design a beam or a column, you also need to determine the maximum and minimum values of axial, shear, and moment that your design will safely support. The problem is, you may not actually know where these locations are. So what are your options?

One option you may choose is to cut the same member at a hundred different locations and perform the same calculations over and over and over. . . . (This task isn't exactly the way I would want to spend a Saturday afternoon.) A better option is to try to formulate a more general set of expressions for internal loads along the length of the bending member.

Generalized internal load equations are typically created for shear and moment at all locations along the length of the member. In the following section, I show you how to create a set of algebraic functions as a function of location in the member.

Defining the critical points

The generalized equations are valid over distinct regions that occur between *critical points*, which are locations on a beam where a change to the shear, the moment, or the relationships between them occurs. Critical points may occur at the following locations:

- ✔ **Start and end of the bending member:** Both ends of the member are considered as critical points.

- ✔ **Support reactions and internal hinges:** All points where support reactions and internal hinges occur are critical points.

- ✔ **All concentrated forces:** All points where concentrated forces occur are critical points.

- ✔ **All concentrated moments:** All points where concentrated moments occur are critical points.

- ✔ **Beginning and end of all distributed loads:** All points where a distributed load starts or stops are considered critical points.

Multiple critical points can (and often do) occur at the same location on your beam.

Consider the simply supported beam shown in Figure 20-7. It's subjected to a linearly distributed load with a maximum intensity of 50 pounds per linear foot at the right end. The slope of this distribution is 50 pounds per linear foot ÷ 20 feet = 2.5 pounds per square foot. Also acting on this beam is an 800-pound concentrated force at 6 feet from Point A and a concentrated moment of 2,000 pound-feet acting at 5 feet from Point B.

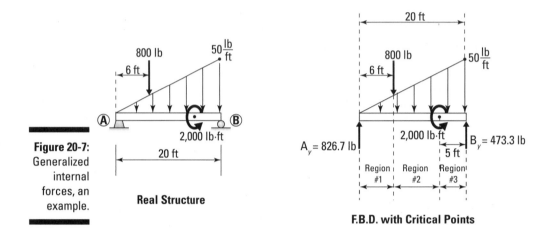

Figure 20-7:
Generalized
internal
forces, an
example.

Real Structure

F.B.D. with Critical Points

Figure 20-7 also shows the F.B.D. of this example, with all loads and moments already included at their appropriate locations on the beam. The support reactions have already been calculated.

The critical points for this structure are indicated by dashed lines on the F.B.D. of Figure 20-7. One critical point occurs at Point A because that's both the left end of the beam and the location of a support reaction. A second critical point occurs at the location of the 800-pound concentrated load. A third critical point occurs at the 2,000-pound-foot concentrated moment. The last critical point occurs at Point B because that is both the right end of the structure and the location of a support reaction.

Establishing the regions of your generalized equations

The generalized equations for internal forces of bending members are valid within specific regions which are shown in the Figure 20-7 example, between those dashed lines that you created for the critical points in the last section. I list them in that figure as Region #1, Region #2, and Region #3. What these regions allow you to do is create a generic F.B.D. that is valid for all points within those regions.

First you need to establish a reference location that remains unchanged for each region. I usually choose to take this reference point as the left end of the structure, but you can choose any point you want — just don't move it! You can then draw F.B.D.s specific for each region, as I've done in Figure 20-8.

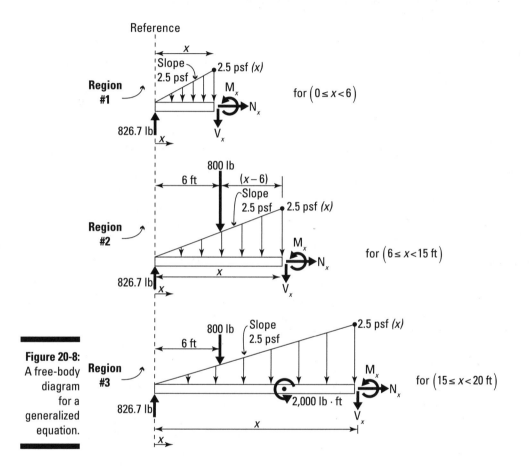

Figure 20-8: A free-body diagram for a generalized equation.

For Region #1, you cut an arbitrary location at an arbitrary distance, x, from your reference. This F.B.D. is valid for all values of x between 0 feet and 6 feet. Whether $x = 1$ foot or $x = 3.697$ feet, this F.B.D. looks exactly the same. The only parameter that changes is the x dimension or the location of the generalized section. Knowing this, you can then employ the equation of equilibrium and write expressions for the shear and moment:

$$+\uparrow \sum \left| \vec{F_y} \right| = 0 \rightarrow 826.7 \text{ lb} - V_x - 0.5 \cdot (2.5x) \cdot (x) = 0$$
$$\Rightarrow V_x = -1.25x^2 + 826.7 \text{ (lb) } (0 \le x < 6)$$

This expression allows you to compute the value of the shear for every location value of x within the region. Similarly, you can compute the generalized moment equations as

$$\zeta + \sum \left| \overrightarrow{M_x} \right| = 0 \rightarrow -826.7 \text{ lb} \cdot \; x + M_x + 0.5 \cdot (2.5x) \cdot (x) \cdot \left(\frac{x}{3} \right) = 0$$

$$\Rightarrow M_x = -0.417 \left(x^3 \right) + 826.7(x)(\text{lb} \cdot \text{ ft}) \; (0 \leq x < 6)$$

Repeating the process of writing equilibrium equations for each F.B.D. of Region #2 and Region #3, you can easily develop the generalized equations. For Region #2,

$$V_x = -1.25x^2 + 26.7(\text{lb}) \; (6 \leq x < 15)$$

$$\zeta + \sum \left| \overrightarrow{M_x} \right| = 0 \rightarrow -826.7 \text{ lb} \cdot \; x + M_x + 0.5 \cdot (2.5x) \cdot (x) \cdot \left(\frac{x}{3} \right) + 800 \text{ lb} \cdot (x-6) = 0$$

$$\Rightarrow M_x = -0.417 \left(x^3 \right) + 26.7(x) \; + 4{,}800(\text{lb} \cdot \text{ ft}) \; (6 \leq x < 15)$$

And for Region #3,

$$V_x = -1.25x^2 + 26.7(\text{lb}) \; (15 \leq x < 20)$$

$$\zeta + \sum \left| \overrightarrow{M_x} \right| = 0 \rightarrow -826.7 \text{ lb} \cdot x + M_x + 0.5 \cdot (2.5x) \cdot (x) \cdot \left(\frac{x}{3} \right)$$

$$-800 \text{ lb} \cdot (x-6) + 2{,}000 \text{ lb} \cdot \text{ft} = 0$$

$$\Rightarrow M_x = -0.417 \left(x^3 \right) + 26.7(x) \; + 2{,}800(\text{lb} \cdot \text{ft}) \; (15 \leq x < 20)$$

These equations represent the generalized equations for internal bending forces. Instead of having to perform a ridiculous number of internal load calculations with a large number of free-body diagrams, you can now create a much smaller number of algebraic expressions that fully define the values of the internal loads within the specific regions.

Discovering other useful tricks from generalized equations

The best part about developing mathematical expressions for the generalized equations (which are actually mathematical functions) is that you can now utilize basic calculus principles to find maximum and minimum values for each of those functions.

Defining the relationship between shear and moment

With calculus, you can also discover an interesting relationship between the function for shear and the function for moment. If you examine the generalized equations for shear and moment for each region, for Region #1 of Figure 20-8, you can compute the first derivative of the internal bending moment M_x.

$$\frac{dM_x}{dx} = \frac{d}{dx}\left(-0.417\left(x^3\right) + 826.7(x)\right) = -1.25\left(x^2\right) + 826.7 \quad (0 \leq x < 6)$$

If you recognize the result of this operation, it's because this derivative actually ends up being the exact same equation as the internal shear function V_x for the same region. You can verify that this relationship is valid for all regions and can thus be defined as $\frac{dM}{dx} = V$ where M is the moment function and V is the shear function for a given region or interval.

The first derivative of the generalized moment function is equal to the generalized shear function over the same interval. Although it may not seem like a big deal at this point, this equation is fundamentally crucial in the advanced study of structural analysis and mechanics/strength of materials classes. This equation also serves as your graphical basis for establishing a shortcut method for creating shear and moment diagrams in the next section.

Calculating maximum and minimum shear and moments with calculus

Another neat feature with the generalized equations is that you can now also calculate the minimum and maximum values for shear and moment over a given region. You've gone to all the work to create nice continuous algebraic functions for shear and moment. You can now apply the principles of calculus that I discuss in Chapter 2 to find the maximum and minimum values of the shear and moment function, as well as their locations.

Plotting a system's internal forces

Another useful result of having these algebraic expressions for the generalized shear and moment equations is that you can now plot their functions on a computer (or by hand) and create a complete shear and moment diagram that shows you every value of the internal shear and moments along the length of the bending member. These shear and moment diagrams are among the more important tools an engineer has available for design.

Creating Shear and Moment Diagrams by Area Calculations

When you compute the generalized internal force equations in the preceding section, you see that those calculations aren't terribly difficult, but they can take a good amount of time (especially if your problem has a lot of critical regions). A faster method is definitely worth investigating.

To help explain a new method based on area calculations, check out Figure 20-9, which contains a simply supported beam with a cantilever at the left end and is loaded as shown. The reactions have already been provided and are indicated on the figure.

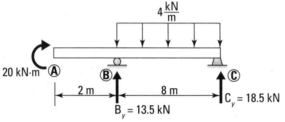

Figure 20-9:
Shear and moment diagram construc- 20 kN·m tion, an example system.

Rules to remember when working with area methods

To use the area methods, load diagrams must be fairly simple and consist of only point loads, concentrated moments, and uniformly distributed loads. Any higher-order distributed loads, and the geometry becomes less than friendly and you're better off just creating the generalized equations.

When you're working with area methods, keep in mind the following:

- ✓ **Construct the framework for the graphs first and align them vertically.** Place the load diagram on top, with all reactions shown as point loads or concentrated moments. Next, place the shear diagram directly below the load diagram and then place the moment diagram directly below that.

- ✓ **Obey the sign convention, and locate all critical points.** You use the sign convention for the left end of the beam, as shown in Figure 20-4. Locate all critical points as I outline in "Defining the critical points" earlier in this chapter.

- ✓ **Finish the shear diagram first.** You must complete the shear diagram before you can start the moment diagram.

- ✓ **Start from the left and work to the right.** This method allows you to use the sign convention as described in the preceding bullet. Place points at $V = 0$ and $M = 0$ on the left end of each diagram.

✔ **The shear and moment diagrams you draw, and the calculations you perform, must compute to zero at the right end of the diagram.** If you don't calculate a zero when you reach the end, you've made an error somewhere along the way. What's worse is that if your shear diagram doesn't equal zero at the end, you're guaranteed to have an incorrect moment diagram.

Constructing the shear diagram

The first diagram that you construct after drawing the loading diagram (or free-body diagram) is always the shear diagram. The shear diagram is built exclusively from the free-body diagram of the entire system, including external applied loads as well as the computed support reactions. If you make a mistake in finding the support reaction calculations, chances are your shear diagram will be incorrect as well.

Using the example in Figure 20-9, you can compute the shear diagram in the following basic steps. Check out Figure 20-10 for a visual.

1. **Starting at Point 1, place a point at V = 0 and then examine the first critical region (between Point 1 and Point 2).**

   If there are no loads acting in this region, the shear remains unchanged, so because the start point is at $V_1 = 0$, the unchanged end point will be at the same value, or $V_2 = 0$. There is a concentrated moment at Point 1, but remember that moments don't affect the shear diagram, so just leave it alone for now.

2. **At Point 2 place a point at a shear value of $V_2 = 0$; draw a line to connect Points 1 and 2.**

   At Point B, a reaction of 13.5 kN is acting upward on the beam. This reaction is the same as a concentrated load, which will cause the shear diagram to jump instantly.

3. **Starting at the value of $V_2 = 0$ at Point 2, add another +13.5 kN to that value (for $V_3 = 13.5$ kN total) and place a point for $V_3 = 13.5$ kN on the graph.**

   This becomes the new value for Point 3.

   Beginning at Point 3, which has a value of $V_3 = +13.5$ kN, you can see that this region is subjected to a uniformly distributed load. The area under this load (or the resultant) is equal to the change in shear value and helps you calculate the shear at Point 4. For this example, the resultant of the distributed load on this region is (–4 kN per meter)(8 meters) = –32 kN. Because this uniform load is acting downward, the resultant must also be acting downward. This fact means that the total change from Point 3 to Point 4 must be –32 kN. The value of shear at point 4 is then (+13.5 kN – 32 kN) $V_4 = -18.5$ kN.

But be careful: The shape of the shear function between Points 3 and 4 is dependent on the shape of the load between those points. The order of the shear function is always one order higher than the load function for the same interval. Thus, a uniform distributed load (with an order of zero) results in a linear (or first order shear function). So a straight line connecting Point 3 and Point 4 is correct. If this load had been linear (first order), the shear function would have been quadratic (or second order).

At this point, you're standing at the end of the beam at Point 4, and sitting on a value of –18.5 kN. But the previous section says that you must end at a value of zero. So what happened? Even though you've reached the end of the beam, remember that there is also a vertical reaction, C_y = +18.5 kN. So the value of Point 5 is equal to the value of Point 4 plus the effect of the concentrated point load. Thus, the shear at Point 5, V_5 = –18.5 kN + 18.5 kN = 0, which means the shear diagram ends on a zero value. Good news!

4. Denote any areas of positive shear with a plus (+) sign inside the region and areas of negative shear with a negative (–) sign.

This notation helps with the moment calculations I discuss in the next section. I also like to shade the areas to make them a bit more visible.

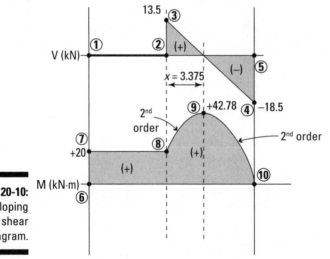

Figure 20-10:
Developing the shear diagram.

5. Look for a secondary critical point that occurs at locations of zero shear.

At any locations where the shear diagram has a value of zero, you need to add a new critical point if there isn't one there already. In this diagram, Point 1, Point 2, and Point 5 are already critical points. However, there is another place where V = 0, somewhere between Point 3 and Point 4. At this location, add a dotted or dashed line and draw it down into the moment diagram, as shown in Figure 20-10. These critical points are locations of maximum or minimum moments.

To calculate the location of this point, you need to employ a bit of geometry. In this example, because the shear function is linear at this location, you can use the concept of proportions or similar triangles to determine the distance of the new critical point from Point 3. Using a variable x to denote this location:

$$\frac{13.5 \text{ kN}}{x} = \frac{18.5 \text{ kN}}{(8 \text{ m} - x)} \Rightarrow x = 3.375 \text{ m}$$

You use this dimension when you calculate the points on the moment curve in the following section. If the functions aren't linear, you need to formulate a generalized shear equation, set the equation equal to zero, and then solve for the x-dimension that satisfies that condition.

Creating the moment diagram

When the shear diagram is complete, you're ready to begin drawing the moment diagram. The moment diagram is based directly off the shear diagram, and all your calculations come from the shear diagram that you previously created. The only exception is the presence of concentrated moments, which cause a jump in the moment diagram. You need to look at the original load diagram to find them!

Using the example earlier in Figure 20-9, follow these steps to create a moment diagram:

1. **Place a point at the left end of the moment diagram at a value of $M_6 = 0$.**

   This becomes Point 6 in this example. At this critical point, the beam also has a concentrated moment in the amount of 20 kN-meters acting clockwise.

   A clockwise moment is a positive moment. This means that Point 7 has a value of $M_7 = (0) + 20$ kN-meters = +20 kN-meters.

2. **Place a new Point 7 at M_7 = +20 kN-meters.**

3. **Compute the change in moment between Point 7 and Point 8 by calculating the area under the shear diagram between the same two critical points (in this case it's the area under the shear diagram between Point 1 and Point 2).**

   Because the shear is zero in this region, the change in moment between Point 7 and Point 8 is (0 kN-meters)(2 meters) = 0 kN-meters. The value at Point 8 is then M_8 = 20 kN-meters + 0 kN-meters = +20 kN-meters. Because there is no shear between Point 7 and Point 8, the moment must remain constant.

4. **Starting at the value of M_8 = +20 kN-meters, compute the change in moment by calculating the area under the shear diagram for the next region.**

$$\triangle M = (0.5)(+13.5\ kN\text{-}m)(3.375\ m) = +22.78\ kN \cdot m$$

The final value of Point 9 is then M_9= +20 kN-meters + 22.78 kN-meters = +42.78 kN-meters. The positive area under the shear diagram means that the moment will increase. So now you have the two endpoints declared, but what does the moment function between them look like?

At this point, you need to do a bit of detective work. The first clue is in the shape of the shear diagram. Remember that moment diagrams are always one order higher than the shear diagram in the same region. So, if the shear diagram is linear (first order), as in this case, the moment diagram must be second order (or one order higher).

But that leaves you with another dilemma. It takes a minimum of three points to define a second-order curve, and you only have two so far! This discrepancy means there are two second-order curves that can possibly fit between Point 8 and Point 9 (see Figure 20-11).

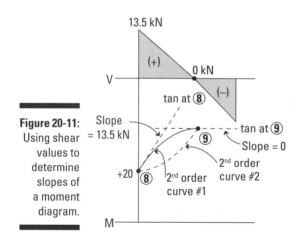

Figure 20-11: Using shear values to determine slopes of a moment diagram.

To deduce which second-order curve actually fits, you need to look at the slope of the moment diagram at each point. Recall that the slope of the moment diagram is equal to the value of the shear at that point. Thus, the slope of the moment diagram at Point 8 is +13.5. The slope at Point 9 is 0. Laying these two slopes on the two possible second-order curves reveals that curve #1 is the right shape and is the only one that is possible. It has the highest positive slope on the left end of the region, where the shear was the highest positive value.

Be careful, though — this second-order curve won't necessarily work for every linear shear function. You have to look at the slopes to make the decision!

5. **Starting at a moment value, M_9 = 42.78 kN-meters, compute the change in moment from Point 9 to Point 10 as the area under the shear diagram in this region.**

   Because the area under the shear curve is negative, you should expect that the change in moment will also be negative.

   $$\triangle M = (0.5)(-18.5 \text{ kN})(8 \text{ m} - 3.375 \text{ m}) = 42.78 \text{ kN} \cdot \text{m}$$

   Thus, the moment value at Point 10 is (M_{10} = +42.78 kN-meters – 42.78 kN-meters) = 0 kN-meters. Hence, the moment diagram ends at a value of zero. You can deduce the shape of the second-order curve in the same manner as in the preceding step. Because the final value was zero, this indicates that the work you did is most likely correct. Be sure to label and shade your positive and negative moment regions as a useful reminder when you're done.

6. **Look at your diagrams and declare the maximum and minimum shear and moment values.**

   By looking at the shear diagram, you can see that the maximum positive shear V_{MAX+} = +13.5 kN at Point B (or Point 3) and the maximum negative shear V_{MAX-} = –18.5 kN at Point D (or Point 4). Similarly, the maximum positive moment, M_{MAX+} = +42.78 kN-meters at the new critical point (or Point 9) and the maximum negative moment, M_{MAX-} = 0 kN at both Point 6 and Point 10. With these values determined, you're ready to begin designing this beam — but I'll save that discussion for another book.

Chapter 21

Working with Frames and Machines

. . . .

In This Chapter

▶ Identifying frames and machines

▶ Blowing apart a structure for analysis

▶ Looking at additional features such as pulleys, sliders, and slotted holes

. . . .

*T*he versatility of frames and machines makes them extremely useful in day-to-day life for the work and convenience that they provide. In fact, the vast majority of structural systems you encounter are classified as either frames or machines and may require you to use slightly different techniques to analyze them than the methods I outline for working with trusses (Chapter 19) and bending members (Chapter 20). The methods I present in those chapters are a whole lot simpler than the methods I present here, so you should apply the rules in this chapter only as a last resort. Though these techniques aren't difficult, they can require a lot of additional work if you're not careful. However, these techniques do allow you to solve problems that you may not otherwise be able to.

In this chapter, I show you how to identify frame and machine systems and then I tell you how to determine if you even need to use these methods in your solution. Finally, I show you the actual procedure for dealing with frame and machine systems and some of the more common components that are added to them.

Identifying a Frame and Machine System

You need to be able to identify systems quickly and efficiently. The following sections can help you to recognize that your system is a frame or machine by discussing the basic properties.

Defining properties of frames and machines

The two types of structural systems that I discuss in this chapter are frames and machines and they have several basic differences:

- ✔ **Frames:** *Frames* are structural systems (such as your house or office building) that are constructed of multiple members, at least one of which must be connected by rigid joints (or joints that don't rotate freely).

- ✔ **Machines:** *Machines* are systems that contain moving parts, such as pulleys, blades, and pistons, that are created to perform a certain task. Hand tools such as pliers, vice grips, and scissors are all examples of simple tools that are classified as machines. Machine systems produce the same additional internal forces in the members as frames, but may or may not have the same rigid joint requirement as a frame. Machines are typically used to transmit forces from one location in the structure to another. For example, a piston assembly in your vehicle's engine receives forces from the combustion process and transmits those forces through the transmission, and into the axles or wheels, causing your vehicle to move.

Aside from these differences, frames and machines also have several properties in common:

- ✔ **Member shapes:** Members do not have to be straight (or linear). They can also be curved or bent into a wide variety of shapes.

- ✔ **Internal forces:** Internal forces on frames and machines have the same type of internal forces as the bending members in Chapter 20 — axial forces, shear forces, and bending moments.

- ✔ **Loads of any type:** Loads can be concentrated loads, distributed loads of any shape and order, or concentrated moments, and they can be applied at any location.

- ✔ **Other items included on the structure:** Frames and machines can also have other items attached to them. In fact, machines almost always have extra tools attached to them, such as pistons and pulleys.

- ✔ **Statically indeterminate:** Frame and machine systems are not necessarily *statically determinate* (or have as many available equations as there are unknown forces on your free-body diagram — more on those later).

Determining static determinacy

In earlier chapters, I show how to find internal forces and support reactions using the equations of equilibrium with three or fewer unknowns on your _free-body diagram_ (or F.B.D.) for two-dimensional problems and six or fewer unknowns for three-dimensional problems. Unfortunately, many structures often are quite complex, and you can encounter many unknown forces on your free-body diagrams. (Check out Part IV for the complete scoop on working with F.B.D.s.)

Consider the F.B.D. of the two-dimensional machine system shown in Figure 21-1, which shows two members pinned together at Point C and supported by pinned supports at Points A and B. A concentrated load is applied on member BC.

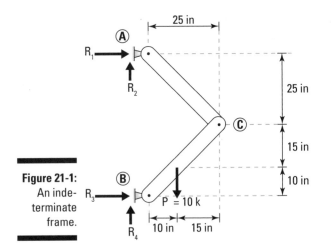

Figure 21-1:
An inde-
terminate
frame.

In this F.B.D., there are two unknown reactions at the pin support at Point A, and two more unknown reactions at the pin support at Point B, for a total of four unknowns on the system. Because you have only three equations of equilibrium available for each F.B.D. of the system, this system is said to have one _degree of indeterminacy_ (or the number of additional unknowns present beyond the number of available equilibrium equations). Because Newton's laws of motion (see Chapter 16) provide you with only three equilibrium equations per F.B.D., you must find additional equations elsewhere.

Using the Blow-It-All-Apart Approach to Solve Frame and Machine Problems

After you've identified the degrees of indeterminacy and confirmed that you dealing with a frame and machine system, the question you face now is exactly how do you solve it?

Breaking it at the hinges

To solve an indeterminate frame and machine system using only statics requires that your system has internal hinges — this will be the place where you cut the system. In many frames and machines, this point is the point where two or more members are connected, such as Point C in the example of Figure 21-1. Figure 21-2 shows the same system broken apart at the hinges.

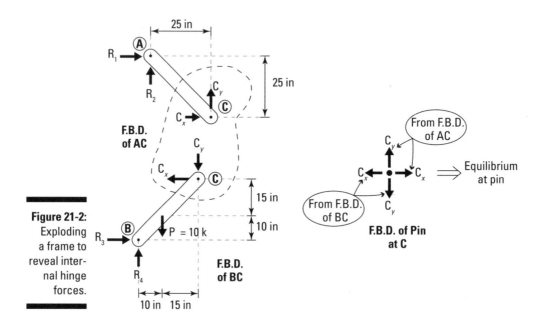

Figure 21-2: Exploding a frame to reveal internal hinge forces.

Next, draw F.B.D.s for each of the separated members. I arbitrarily assume the pinned forces acting on member AC are such that C_x is acting to the right, and C_y is acting upward on member AC. Member BC has the unknown support reactions and the concentrated load **P** and forces on the pin at Point C.

Because the pin at Point C itself must also be in equilibrium, and because I already assume that the C_x component is acting to the right on the F.B.D. of member AB, the C_x pin force on member BC must be in the opposite direction. Likewise, the pin force C_y must be assumed to be acting downward on member BC because I already assumed that its counterpart was acting upward on AC.

Don't be alarmed if you perform your equilibrium calculations and discover that your forces have a negative value. Just as with other equilibrium calculations, if you assume the wrong direction for the forces on your F.B.D., the signs of your calculations will always tell you. As long as you ensure equilibrium of the pin with your assumed directions (by making sure that all forces on the pin are balanced), your calculations will still work.

Knowing where to start solving frame and machine problems

For each two-dimensional F.B.D. you're able to draw properly, you're allowed to use up to three equilibrium equations — two translational equations and one rotational equation.

In the earlier example of Figure 21-2, you now have a total of six unknown forces on your two F.B.D.s: the four reaction forces R_1, R_2, R_3, and R_4 as well as the two unknown hinge forces C_x and C_y. Because you have two members on two separate F.B.D.s, you now have 2 members × 3 equations per member = 6 total equilibrium equations to work with. This means you now have six unknown forces and six equations of equilibrium which means you have a statically determinate set of free-body diagrams, or a solvable system. Now you just have to figure out where to start.

Starting on a member with a load

For most frames and machines, you typically want to start writing your equilibrium equations for the member that has a known applied load on it. For the example of Figure 21-2, I start with member BC because it has a known load value acting on it: 10 kip downwards. The two translational equilibrium equations that you write are

$$\pm\sum\left|F_x\right|^{\text{BC}}=0\rightarrow R_3-C_x=0$$
$$+\uparrow\sum\left|F_y\right|^{\text{BC}}=0\rightarrow R_4-10\text{ k}-C_y=0$$

Solving for the unknown hinge forces as fast as you can

The trick for solving most frame and machine problems is in determining the unknown hinge forces because they provide you with the extra information to use on all the other attached members. For the example of Figure 21-2, your goal is to get expressions for the hinge forces C_x and C_y.

You still have a rotational equilibrium equation remaining to write for member BC, so you want to write this equation with Point B as your reference. (After all, you don't want to have to deal with the reactions R_3 and R_4 yet, so sum moments at their point of application, and they vanish from the moment equations.)

$$\zeta+\sum\left|M_B\right|^{\text{BC}}=0\rightarrow-\left(10\text{ k}\right)\cdot\left(10\text{ in}\right)+C_x\cdot\left(25\text{ in}\right)-C_y\cdot\left(25\text{ in}\right)$$
$$\Rightarrow C_x-C_y=4\text{ k}\cdot\text{in}$$

This produces a handy relationship between the hinge forces C_x and C_y. You then switch to each of the remaining F.B.D.s and perform the similar equilibrium calculations. First, you write the translation equilibrium equations for member AC:

$$\pm\sum\left|F_x\right|^{AC} = 0 \rightarrow R_1 + C_x = 0$$
$$+\uparrow\sum\left|F_y\right|^{AC} = 0 \rightarrow R_2 + C_y = 0$$

The four translational equilibrium equations that you've written for members AC and BC are all related to one of the hinge forces at Point C. Finally, I sum moments at Point A on member AC to produce another equation for the pin forces at Point C.

$$\circlearrowleft+\sum\left|M_A\right|^{AC} = 0 \rightarrow C_x \cdot \left(25 \text{ in}\right) + C_y \cdot \left(25 \text{ in}\right) = 0$$
$$\Rightarrow C_x = -C_y$$

You now have a second relationship between the same hinge forces C_x and C_y. Substituting the rotational equilibrium equations for member AC into the rotational equilibrium equation for BC:

$$C_x - (-C_x) = 4 \text{ k}$$
$$2C_x = 4 \text{ k}$$
$$C_x = +2 \text{ k}$$

Now that you know the horizontal force C_x, you can substitute into either equation and solve for the remaining hinge force C_y:

$$2 \text{ k} - C_y = 4 \text{ k}$$
$$C_y = -2 \text{ k}$$

After you know the hinge forces, you can then substitute into the remaining equilibrium equations and solve for the support reactions:

- ✔ **For R_1:** $R_1 + 2 \text{ k} = 0$, so $R_1 = -2 \text{ k}$
- ✔ **For R_2:** $R_2 + (-2 \text{ k}) = 0$, so $R_2 = 2 \text{ k}$
- ✔ **For R_3:** $R_3 - 2 \text{ k} = 0$, so $R_3 = 2 \text{ k}$
- ✔ **For R_4:** $R_4 - 10 \text{ k} - (-2 \text{ k}) = 0$, so $R_4 = 8 \text{ k}$

To solve statics problems involving frame and machine problems and other indeterminately hinged structures, you must find the hinge forces first. If more than one hinge is on a structure, you must solve for the forces on each hinge.

Considering Other Useful Approaches to Common Frame and Machine Problems

While the concept of "blow-it-all-apart" is a pretty sound technique for solving frame and machine problems, you quickly discover that deciphering the forces that you're dealing with is sometimes a bit harder. In this section, I give you a bit more insight in handling some of the more common situations that you might encounter.

When more than two members meet at an internal hinge

Some frames and machines have more than two members that are connected at the same hinge location. Consider the example shown in Figure 21-3, which has three axial members connected to the same internal hinge at Point D.

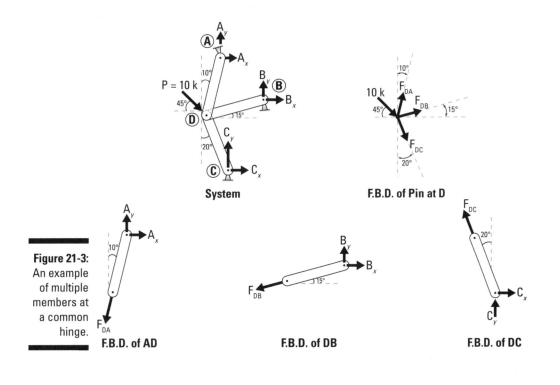

Figure 21-3: An example of multiple members at a common hinge.

The first figure shows the external pinned support conditions acting at Points A, B, and C, for a total of six unknown forces acting on the combined system. With only three equilibrium equations available to solve for six unknowns, you now know that this structure is statically indeterminate to three degrees. This means that you need an additional three equations to solve for the reaction forces. Yikes!

Your first inclination may be to automatically assume that this system meets all the criteria to be solved as a truss with the method of joints (see Chapter 19 for more information on trusses). After all, each member is pinned at the end, each member is axially loaded, and the point load is concentrated and applied at a hinge point. Clearly, this is a truss, right?

In order for the method of joints to work, you can have a maximum of two unknown forces on any given joint at the time you're solving it. This rule means you can't use the same truss solution techniques to solve this problem.

In this example, you actually have three members (with three internal axial forces). At this point, you're left with only one option: using the frame and machine solution techniques of this chapter.

As you did in the previous example, you blow apart the system at all hinge locations and draw free-body diagrams of each member. The F.B.D. for member AD consists of the pin support reactions A_y and A_x and the force of the hinge at Point D on the member F_{DA}, for a total of three unknown forces.

Similarly, member DB has two unknown support reactions B_x and B_y, and the unknown pin force F_{DB}, for a total of three unknowns for this member. Finally, member DC also has two unknown support reactions, C_x and C_y, and the unknown pinned force F_{DC}, for a total of three more unknowns for this member.

In total, this means that there are three members, with three unknowns per member, for a total of nine unknown forces acting on the exploded frame system. But you have three equations per F.B.D. that you created, so that means that you now also have nine equilibrium equations to work with. Nine unknowns and nine equilibrium equations means that this system is now statically determinate and thus solvable!

Dealing with pesky pulley problems

A *pulley* is a mechanical apparatus consisting of a round object attached to a shaft that allows the object to rotate. A belt or a rope is then wrapped around part of the pulley and is used to transmit a force from one location to another. This shaft is then attached to bearings (making it a pin support), which is then attached to an external structure or support.

A pulley serves two basic purposes: It provides a convenient change of angle for a force, and when combined with multiple other pulleys, they can help you lift a much heavier object than you might otherwise be capable of lifting. A pulley is also a very convenient mechanism for transmitting forces between mechanical parts. Just look under the hood of your car — several different parts of your engine are connected by a variety of pulleys and belts.

Changing force direction with pulleys

Pulleys are used to change the direction of force. Figure 21-4 shows a person pulling on a rope wrapped around a pulley, lifting a weight W.

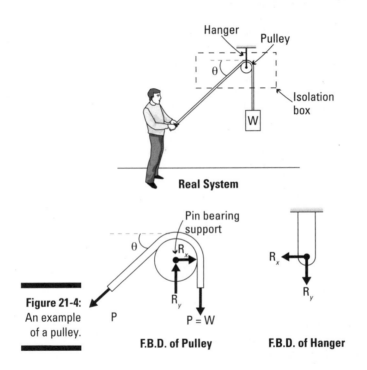

Hanger

Pulley

θ

Isolation box

W

Real System

Pin bearing support

θ

R_x

R_x

R_y

R_y

P

$P = W$

Figure 21-4: An example of a pulley.

F.B.D. of Pulley

F.B.D. of Hanger

In many statics problems, pulleys are often considered as *frictionless*, meaning, the motion of the pulley, or the wrapping of the belt around the pulley, does not provide any resistance to the person pulling on the rope. (Don't worry, in Chapter 24, I discuss more about friction problems.)

Like other machine F.B.D.s, working with pulleys requires you to break the system apart into individual pieces. For pulley problems, you first draw an isolation box (see Chapter 14) around each individual pulley and cut any necessary ropes or belts — belts and ropes are always axial tension members, so you always know the direction of those forces on the F.B.D. Be sure to also include the support reactions of the pulley on the F.B.D. You then treat the pulley F.B.D. as any other exploded frame or machine part.

You get three additional equilibrium equations for each pulley F.B.D. you can create from a system of pulleys.

Creating mechanical advantage with pulleys

Interconnecting multiple pulleys and wrapping the same rope multiple times around the assembly is known as a *block-and-tackle* assembly. Block-and-tackle assemblies are used to increase the amount of force lifted on one end of a rope for a given applied load at the other end. Consider the three-pulley system of Figure 21-5, which shows a rope passing around Pulley A, wrapping around Pulley C and over Pulley B, and connecting to the middle of Pulley C again. Suppose you're interested in calculating the force required to lift a 900-pound weight suspended from Pulley C. For this example, I assume that each of the pulleys is frictionless.

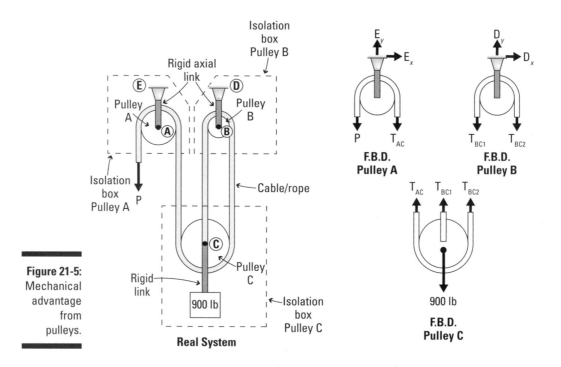

Figure 21-5:
Mechanical advantage from pulleys.

As with other machine problems, your first step is to explode the system into as many pieces as possible and then draw your F.B.D.s. In this example, I include the rigid links and supports on my F.B.D. for Pulleys A and B. If I had chosen to, I could have drawn a separate F.B.D. for each of the rigid links as well, but because I'm not interested in finding their internal forces, and because they're axial members connected to a support (I recognize this because the links are pinned at both ends and don't have any loads acting directly on them), their F.B.D.s would provide little useful information with regards to calculating the applied load P. As long as I follow my F.B.D. checklist (see Chapter 13), I can actually cut this system in any way I want. I just have to make sure to get all the forces, support reactions, internal axial forces from the cables, and self weight (if there were any) applied at the proper locations on the F.B.D. For this problem, I have three unique free-body diagrams to consider, one for each pulley that I removed from the system (refer to Figure 21-5).

For multiple pulley problems, you want to start with the pulley that has a known load value attached to it. In this case, I'll start with Pulley C, which is supporting 900 pounds.

In drawing the isolation box around Pulley C, I had to cut three cables, which means I have to include one internal axial tension force for each of the cables that were cut — Pulley A and its link to its support, Pulley B and its link to its support, and Pulley C and its support weight of 900 pounds. Applying the equations of equilibrium in the y-direction:

$$+\uparrow \sum \left|F_y\right| = 0 \rightarrow T_{AC} + T_{BC1} + T_{BC2} - 900 \text{ lb} = 0$$

This equation has three unknown tension forces currently acting on it. But, because this F.B.D. contains a *single rope* wrapping multiple times around multiple *frictionless* pulleys, I know that the tension in each of the rope segments must be equal. That is, $T_{AC} = T_{BC1} = T_{BC2} = T$, where T is the tension in each rope segment.

Rewriting the previous equation, I can now solve for the tension T in the rope:

$T + T + T - 900 \text{ lb} = 0$

$T = 300 \text{ lb}$

Next, to determine the amount of force applied I need to look at the F.B.D. for Pulley A. This F.B.D. reveals that only a single rope is acting around Pulley A. So now I know that the tension in the rope on one side of the frictionless Pulley A must be equal to the tension on the other side. This means that:

$P = T_{AC} = 300 \text{ lb}$

The force required to hold the 900-pound force on this tackle assembly in equilibrium is 300 pounds.

Because of the mechanical advantage of the way this system is set up, I only need one-third (or 33 percent) of the load in the rope to balance the applied weight. In fact, the more times you wrap the same rope around the same system of pulleys, the less weight will be required in the rope to balance a suspended load. Talk about advantage!

Tackling Complex and Unique Assemblies on Machine Problems

More-complex machines may include unique attachments that provide you with a specific functionality. Some examples of these assemblies that you use in statics are pistons and slotted connections.

Pistons and slider assemblies

The common piston is a tool that you may find in your car engine or an industrial stamping machine (see Figure 21-6).

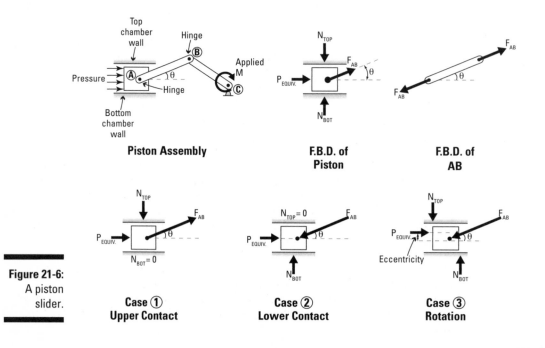

Figure 21-6: A piston slider.

In your car, the piston experiences a pressure on one edge (or face) due to the combustion of gasoline in the engine. This pressure resultant is a single force, P_{EQUIV}, that is applied to the piston at the midpoint (assuming uniformly distributed pressure). The natural response of the piston is to move in the direction of the applied force, which then exerts a force into the bar AB, which can be oriented at any given direction (in this case, at an angle θ).

If the forces are all concentric (or concurrent), ask yourself this: With just these two forces on the F.B.D., is the piston in equilibrium? Without either of the normal forces N_{BOT} or N_{TOP}, the piston wouldn't be in balance in the perpendicular (or vertical) direction for this example. That means that you'd need at least one contact force of the piston against the sidewall in at least one direction. But which one needs to be included?

In free-sliding assemblies, such as pistons and slider mechanisms, there are actually multiple cases that you need or want to investigate. For the simplest cases, you need to consider the events of the first two of the following cases. For more complex cases, something like the third one would be a possibility as well.

- ✔ **Case 1:** If the force F_{AB} were to pull on the piston in the direction shown (up and to the right), the response of the piston would be to move upward and to push against the top edge of the chamber. But as it makes contact with the top edge, it loses contact with the bottom edge (making N_{BOT} = 0).

- ✔ **Case 2:** If the force F_{AB} reverses direction such that it is now pushing on the piston in the opposite direction, the piston will want to move down in response. As it moves down, it makes contact with the bottom wall and loses contact with the top edge (making N_{TOP} = 0).

- ✔ **Case 3:** If the forces become nonconcurrent, which can occur if the location of P_{EQUIV} varies (such as would happen with an improper combustion firing) and develops an eccentricity, or if the applied pressure isn't uniformly distributed, the resultant location would no longer be at the midpoint. The force acting at an eccentricity from the point of application of force F_{AB} would actually cause a rotation of the piston within the chamber and produce separate normal forces on both the top and bottom walls at uniquely different locations. These forces would then partially act as a couple to resist the rotation due to the eccentric applied load.

The major issue with these types of problems is that you often don't know which case to start with. This situation means that you have to choose a case by making a guess and then verify that the numerical values and their signs from your calculations all logically make sense. (For more on this idea, turn to Chapter 24.)

Slotted holes and unidirectional pins

Slotted holes are another feature commonly encountered in engineering designs. A *slotted hole* is an elongated hole (often several inches in length) that allows one piece to move relative to another in just one direction while maintaining strict contact in the other.

The purpose of the slot is to remove the restraint in a given direction. Without restraint, support reactions can't develop.

Consider the example shown in Figure 21-7. This apparatus contains a slotted hole that connects two members of a frame assembly. The analysis of this is the same as if the members were connected with an internal hinge, with one exception. Notice that on the F.B.D. of member ACD, the horizontal pinned force C_x is no longer present because the restraint in that direction has been removed by the slot. Likewise, the opposite force on member BC is also missing. Force C_y can be assumed to be acting in any given direction on ACD just as before. **Remember:** The opposite force direction must be applied on member BC.

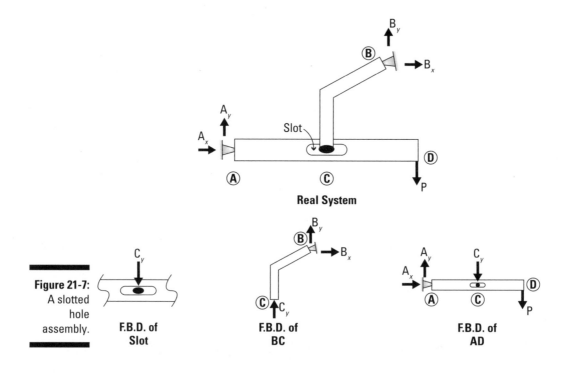

Figure 21-7:
A slotted hole assembly.

Chapter 22

A Different Kind of Axial System: Cable Systems

C able systems are common structures in engineering that are popular because of their relatively lightweight construction and their aesthetic beauty. When someone says, "Picture a bridge in your mind," many people in the United States think of the Golden Gate Bridge in San Francisco, renowned for its glowing red shape and long, slender cable system.

Although you still must observe all the rules of equilibrium I discuss in Part V, cable systems provide a unique set of challenges in that the forces in the cable structures are all dependent on the geometry of the cable system. For the same applied loads, you can get completely different geometrical behaviors!

In this chapter, I explain the three major categories of flexible cable systems (concentrated loaded, uniformly loaded, and catenary) and the properties that make cable systems unique. I show you how to calculate the tension and deflection (known as *sag*) in cables for each of the different types of system and introduce a shortcut method known as the beam analogy (but try to resist jumping directly to that discussion).

Defining Nonlinear Structural Behavior

In Chapter 19, I show you trusses, which are systems of multiple members (objects) that are *axially loaded* (or members whose internal forces are all acting in the direction of their longitudinal axis). Another type of important axially loaded-only member is the flexible cable, which I cover in this chapter. *Flexible cables* are commonly used in a wide variety of applications, including power and telephone lines, aerial trams, and suspension bridges.

The internal forces in a cable system depend significantly on the following factors:

- ✔ **Cable tension force:** *Cable tension force* is the internal axial tension of the cable. Cables and ropes can only support axial tension — they can't support compression, shear, or moments (which I cover in Chapter 20).

- ✔ **Sag:** *Sag* is a measure of the displacements of a cable system and directly affects the internal forces of the cable. Tension in cables can change dramatically for the same given applied loads with a simple variation in the sag of the cable. The sag for all cable systems varies with position, and you'll never see a suspended cable that is entirely horizontal.

- ✔ **Geometry:** *Geometry* includes factors such as the span of the cable system, the support locations, and the elevations of the cable supports and applied loads.

Cables are assumed to be axial tension-only members and have negligible resistance to bending. End supports of cables are always assumed to be *pinned supports* (see Chapter 13) which are free to rotate but can't move (or *translate*) otherwise. You can't have a roller support on a cable system.

In general, you can divide cable systems into three major classifications:

- ✔ **Concentrated load systems:** *Concentrated load systems* are cable systems with *point loads* (or concentrated forces) acting on them. You can apply multiple concentrated loads, but they must be spread out and not be continuous or distributed loads. (Flip to Chapters 9 and 10 for more on concentrated and distributed loads, respectively.)

- ✔ **Uniform load systems:** *Uniformly loaded cable systems* are those systems that are loaded by a constant, uniformly distributed load acting over a horizontal length. A uniformly loaded cable system is sometimes referred to as a *parabolic cable system*.

- ✔ **Catenary systems:** A *catenary system* is a cable system that is deflecting under its own *self weight* (the force created by gravity's effects on the system's mass) or is subjected to a load that acts along the length of the cable itself (as opposed to a horizontal dimension).

Distinguishing among Types of Flexible Cable Systems

Loads on cables are typically vertical and are applied as either concentrated loads or distributed loads. Self weight of cable systems is often assumed to be negligible because it's usually significantly smaller in comparison to applied concentrated and uniform loads. The load type determines the overall shape of the geometry and is the primary factor in choosing a solution technique. I cover three major types of cable systems in the coming sections.

Recognizing cables under concentrated loads

A cable system subjected to concentrated loads deflects into a shape that resembles a series of straight line segments. The tension in each of these segments may have a different *magnitude* (or the size of the internal cable tension force) for a given *cable segment* (section of cable between concentrated loads). Concentrated systems are the easiest to work with because the tension remains constant over a given cable segment and is directly related to the angles of the cable segments, which are based on the sag of the system at the point loads. Figure 22-1 shows a cable system subjected to concentrated point loads.

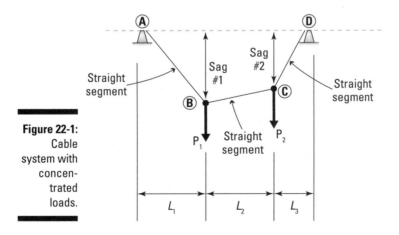

Figure 22-1:
Cable system with concentrated loads.

You can use simple geometry to calculate the angles of the cable segments, or you can use proportion triangles (which I cover in Chapter 5) to set up relationships. Cable systems lend themselves very well to proportion triangle calculations because you almost always know the vertical dimensions (sag) and the horizontal dimension (cable segment length or distance to point loads). I explain more in "Solving for Tension in Flexible Cables" later in the chapter.

Picking out parabolic cable systems

A *parabolic cable system* develops when a uniform load is applied horizontally along the full length of the cable. An example of this type of loading may be the result of roadway decks on suspension bridges. Figure 22-2 shows a parabolic cable system subjected to uniform loads.

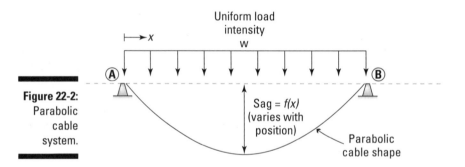

Figure 22-2:
Parabolic
cable
system.

The deflected shape of a cable subjected to a uniform load is a continually curved or parabolic cable-system shape. Suppose you have a parabolic cable system that is loaded with 20 pounds per linear foot (or plf) for a cable that is 15 feet long and tied to supports that are separated by a distance of 10 feet. You can determine the total uniform load (or the resultant) acting on this system with the following equation:

Total Load = (20 plf) · (10 ft) = 200 lb

The parabolic system is only dependent on the intensity of the load and the horizontal projection distance over which it acts.

Identifying catenary cable systems

As I note earlier in the chapter, a catenary cable system deflects under a load along the length of the cable such as ice on power lines or even the weight of the cable itself. Figure 22-3 shows a catenary cable system subjected to uniform loads.

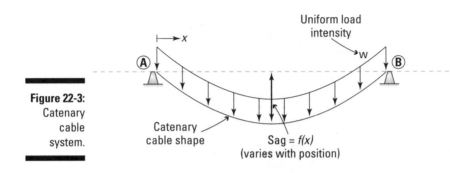

Uniform load intensity

w

Ⓐ Ⓑ

Catenary cable shape

Sag = *f(x)*
(varies with position)

Figure 22-3:
Catenary
cable
system.

Suppose you have a catenary system loaded with 20 pounds per linear foot (or plf) for a cable that is 15 feet long and tied to supports that are 10 feet apart. You can find the total uniform load acting on this catenary system with the following formula:

Total Load = (20 plf) · (15 ft) = 300 lb

You can see that this catenary system has a significantly higher load because the resultant load (or total load) from this system is directly related to the length of the cable itself.

This equation looks strikingly similar to the one for parabolic cable systems in the preceding section, but it uses some different terms. Be sure you plug in the correct numbers for horizontal span or cable length depending on the type of system.

For very small-intensity distributed loads (or even very small sag amounts), catenary systems are basically the same as parabolic cable systems.

Solving for Tension in Flexible Cables

After you know how to identify the three major types of flexible cable systems (see the preceding section), you can calculate the sag and cable tension for each type.

One of the problems you face when working with cable structures is that you often don't know the sag until after the load has been applied. And if you don't know the sag, you may find it difficult to determine the maximum tension in the cable. And if you don't know the maximum tension in the cable, you can't actually design the cable to hold the loads in equilibrium. And if you can't design the cable, you can't predict the sag. . . . And hence, you see the major problems with cable systems: Where do you start?

The forces in the cables become directly dependent on the geometry of the system. As the geometry changes, the forces in the systems change. To get around this issue, you generally assume one parameter, such as the sag or the cable tension values, and then solve the problem for the other value.

Because cable loads are always assumed to be vertical, the horizontal components at the reactions are constant and in opposite directions to each other. This setup means that the horizontal component at every location along a cable is also constant.

Concentrated load systems

You apply a method similar to the method of joints for trusses (which I explain in Chapter 19) at every concentrated load location, treating each cable segment on either side of the load as an individual two force member and drawing a free-body diagram (F.B.D.; see Part IV) of each "joint" (or concentrated load). You can compute the tensions in the cable because only the two unknown force vectors are acting at that point.

Consider the two cases of Figure 22-4, where two cable systems with the same span support the same 100-pound load in the middle. The only difference is that Case 1 has a sag of 6 inches, and Case 2 has a sag of 8 inches.

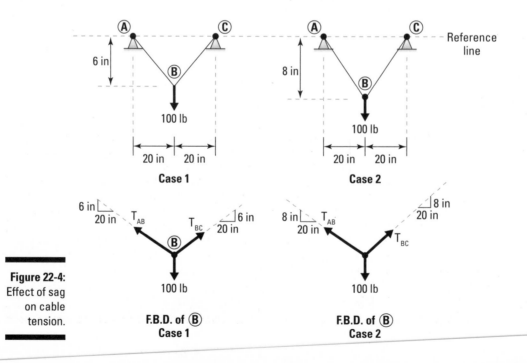

Figure 22-4:
Effect of sag on cable tension.

Using a simple F.B.D., you can see that two cable forces T_{AB} and T_{BC} are acting at Point B. Because the system in this example is symmetrical, you can take advantage of symmetry to note that

$$T_{AB} = T_{BC}$$

Equilibrium (see Part V) in the vertical direction then yields the following equation for the cable tension in Case 1:

$$+\uparrow \sum \left| \vec{F_y} \right| = \frac{6 \text{ in}}{20.88 \text{ in}}(T) + \frac{6 \text{ in}}{20.88 \text{ in}}(T) - 100 \text{ lbs} = 0 \rightarrow T = 174 \text{ lbs}$$

Note that the 20.88 inches value is the hypotenuse of the proportion triangle for the forces in the cable segments. (Chapter 5 gives you the lowdown on proportion triangles.) For Case 2 (which has a different hypotenuse value of 21.54 inches now), you have a similar equation:

$$+\uparrow \sum \left| \vec{F_y} \right| = \frac{8 \text{ in}}{21.54 \text{ in}}(T) + \frac{8 \text{ in}}{21.54 \text{ in}}(T) - 100 \text{ lbs} = 0 \rightarrow T = 135 \text{ lbs}$$

As you can see from this example, a small change in the amount of sag can make a significant difference in the tension in the cable.

Larger cable tensions usually occur near concentrated loads. Large tension loads can also occur at locations with very small sags. If you don't believe me, try it out in the nearby "Relating tension to sag" sidebar.

Relating tension to sag

To see the relationship between tension and sag, hold a small piece of string between your fingers at a set distance apart, allowing it to have a moderate amount of sag. Hang a small weight from this cable and observe its final resting position. Now, maintaining your hold on the string at the same locations (which keeps the support reactions the same), gradually apply tension to the cable and observe what happens. A slight increase in the tension of the string causes the sag to decrease, which then causes the cable to lift the supported load. Repeat this process, and you notice that the sag continues to decrease as you increase the tension.

However, you quickly discover that to reduce the sag by the same amount each time, you must apply more and more tension (by pulling harder) on the cable. As you get to really small amounts of sag, you may actually have to apply a force that causes the string to break. You may also see that you can never fully remove all of the sag from the cable, because a small amount of sag is necessary to develop the vertical component of the tension in the cable, which is what keeps the supported load in equilibrium.

Parabolic cable systems

The first step to solving parabolic cable problems is to locate an origin for a coordinate system at the location of maximum sag. At this point, you must align the *x*-axis with the horizontal direction and the *y*-axis with the vertical.

If you remember your basic calculus, you know that the tangent to a curve at a point has a slope of zero at a location of maximum or minimum value. For cables, this means that the axial force is acting horizontally at the point of maximum sag, which means that the tensile force T_o is horizontal, and the vertical component of the cable tension is zero at that location. In calculus terms, this information means you can create an expression:

$$\frac{d^2 y}{dx^2} = \frac{w}{T_o}$$

which is the differential equation that governs the behavior of flexible parabolic cables.

The reason for this shift of the coordinate system location is that the shape of the cable becomes symmetric at the point of maximum sag. This symmetry doesn't mean that the dimensions L_A and L_B are necessarily the same value, but rather that the overall parabolic shape is symmetrical about that point. Based on this new origin, you can define $y = 0$ at $x = 0$ (this setup is called a *boundary condition*). By integrating the governing differential equation and applying a little bit of algebra to the known boundary condition, you can produce the expression for the deflected shape of the cable:

$$y = \frac{wx^2}{2T_o}$$

So with this expression, all you need to know is the vertical load applied, *w*, and the horizontal component of the tension in the cable, T_o, and you can determine the amount of sag *y* at any point *x* measured from the origin.

Figure 22-5a shows a parabolic cable with a uniform load of intensity acting on a horizontal length.

The F.B.D. of segment BC in Figure 22-5b shows that for a given segment, the horizontal component of the tension at all points must be equal to T_o, or

$$\pm \sum \left| \overline{F_x} \right| = 0 \rightarrow -T_o + T\cos\theta = 0 \rightarrow T_o = T\cos\theta = T_x$$

As you move further from the origin location, the horizontal component remains the same, but the vertical component increases to its maximum value at the support locations. Thus, the maximum magnitude of the tension occurs at the support location.

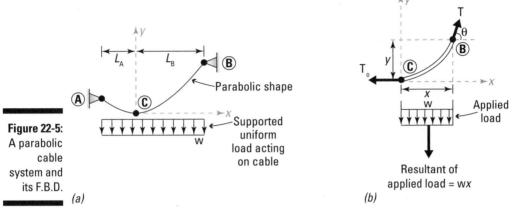

Figure 22-5:
A parabolic cable system and its F.B.D.

(a)

Parabolic shape

Supported uniform load acting on cable

(b)

Resultant of applied load = wx

Applied load

TIP

This characteristic also means that the horizontal component T_0 must be the same magnitude as the horizontal pin support reactions (if the applied loads are all vertical).

Calculating tension when you know sag

You can use a system's known sag to figure out how much tension it's supporting. Consider the suspension bridge in Figure 22-6a, which has a span between the towers of 500 feet (assumed to be pinned at the towers), and a maximum sag of 35 feet at the midpoint. A deck load of 1 kip per foot is applied over the horizontal length of 500 feet. To begin your analysis, you draw an F.B.D. (as shown in Figure 22-6b) of the cable between the towers and treat their supports as pinned at both ends.

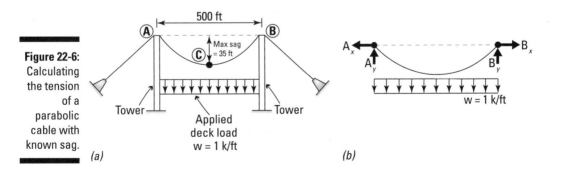

Figure 22-6:
Calculating the tension of a parabolic cable with known sag.

(a)

500 ft

Max sag = 35 ft

Tower

Tower

Applied deck load
w = 1 k/ft

(b)

w = 1 k/ft

The first step to find the horizontal tension component (T_0) in the cable is to locate the coordinate system at the point of maximum sag at a distance of 250 feet from both Point A and Point B, measured horizontally. Thus, at a distance of $x = 250$ feet from the origin, $y = 35$ feet (which is the maximum sag).

Rearranging the basic parabolic cable equation and solving for horizontal cable component T_o gives you the following equation:

$$T_o = \frac{wx^2}{2y} = \frac{(1\text{ k/ft})\cdot(250\text{ ft})^2}{2(35\text{ ft})} = 893\text{ k}$$

The horizontal component has the same magnitude as the support reactions, A_x and A_y, so

$$T_o = A_x = B_x = 893\text{ k}$$

If you sum moments at Point A, you can then solve for B_y, the vertical reaction at Point B:

$$\circlearrowleft + \sum |\overline{M_A}| = 0 \rightarrow \cancel{A_x(0)} + \cancel{A_y(0)} + \cancel{B_x(0)} + B_y(500\text{ ft}) - (1\text{ klf})(500\text{ ft})\left(\frac{500\text{ ft}}{2}\right) = 0$$
$$\Rightarrow B_y = 250\text{ k}$$

Now that the vertical and horizontal components of the support reactions have been determined, you can compute the maximum magnitude of the tension in the cable:

$$T_{MAX} = \sqrt{(B_x)^2 + (B_y)^2} = \sqrt{(893\text{ k})^2 + (250\text{ k})^2} = 927\text{ k}$$

Calculating sag when tension is known

The solution process for calculating sag from tension is basically the reverse of the tension-from-sag calculation in the preceding section. Suppose that for the suspension bridge shown in Figure 22-6a, the design engineer tells you that you're limited to a maximum of 750 kip of tension in the cable. You can show that the vertical component of the reaction remains unchanged at 250 kip. (See the preceding section for this calculation.)

Rearranging the magnitude equation, you can compute the horizontal component of the cable tension as

$$T_o = B_x = \sqrt{(T_{MAX})^2 - (B_y)^2} = \sqrt{(750\text{ k})^2 - (250\text{ k})^2} = 707\text{ k}$$

Recall that the horizontal component of the cable tension is the same value as the horizontal reactions (assuming that all loads are vertical, as they are in this example). Utilizing the boundary conditions for this problem at $x = 250$ ft, $y =$ maximum sag, you can create the following equation:

$$\text{sag}_{MAX} = \frac{wx^2}{2T_o} = \frac{(1\text{ k/ft})\cdot(250\text{ ft})^2}{2(707\text{ k})} = 44.2\text{ ft}$$

By comparing the results in the examples in this section, you can see that by allowing the cable to sag almost 10 additional feet, you can reduce its tension from 927 kip to 707 kip (or by roughly 25 percent).

Catenary cable systems

Catenary cable problems are a little more mathematically challenging because the uniform load applied is now a function of the cable length. The more sag a cable system has, the more cable length is present to carry the load.

The derivation for a catenary problem is very similar to the derivation you use for parabolic cables (see "Parabolic cable systems" earlier in the chapter). You still must define your coordinate system such that the origin is at the location of maximum sag and the horizontal component of tension T_o is equal to the cable tension at that point.

Figure 22-7a shows the same cable as Figure 22-5, except now the load is applied per length of cable, not on a horizontal basis. To clarify this difference, I've changed the applied load intensity from a w symbol to a μ symbol. Also, the length over which the load acts is now the arc length of the cable μs. The origin is placed at the location of maximum sag, (at Point C) as shown in Figure 22-7b.

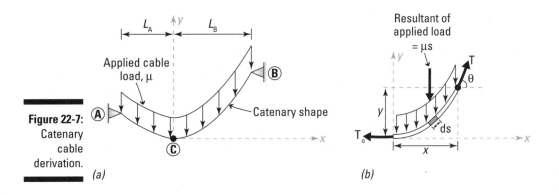

Figure 22-7: Catenary cable derivation.

(a)   *(b)*

Applying a little calculus gives you the following governing differential equation for the catenary cable:

$$\frac{d^2y}{dx^2} = \frac{\mu}{T_o} \cdot \left(\frac{ds}{dx}\right)$$

By comparing this equation to the parabolic version, the load of the parabolic system equations, w(dx), has been replaced with μ(ds). The major mathematical difference is that the incremental arc length term ds is now also a function of both the sag (y) and position (x).

Solving the differential equation and applying a bit of algebra, you can determine that the sag of a catenary cable can be expressed by the following equation:

$$y = \frac{T_o}{\mu} \cdot \left(\cosh\left(\frac{\mu x}{T_o} \right) - 1 \right)$$

where the *cosh* term (which you probably remember from trigonometry) represents the hyperbolic cosine trig function or

$$\cosh(a) = \frac{e^a + e^{-a}}{2}$$

In the catenary equation, the *a* term in the cosh definition is

$$a = \frac{\mu x}{T_o}$$

Finally, the magnitude of the tension *T* at any point can be shown to be related to the sag *y* at that point:

$$T = T_o + \mu y$$

If the horizontal component of the tension T_o in Figure 22-6 is the same 893 kip, the load applied to the cable is still 1 kip per foot, and the span between supports is still 500 feet, you can now compute the sag of a catenary system at a distance of $x = 250$ feet from the new origin with the following equation:

$$y = \frac{T_o}{\mu} \cdot \left(\frac{e^{(a)} + e^{-(a)}}{2} - 1 \right)$$

where the parameter *a* is given by

$$a = \frac{1\,\text{klf} \cdot 250\,\text{ft}}{893\,\text{k}} = 0.280$$

which lets you then solve for the sag in the catenary cable:

$$y = \frac{893\,\text{k}}{1\,\text{klf}} \cdot \left(\frac{e^{(0.28)} + e^{-(0.28)}}{2} - 1 \right) = 35.23\,\text{ft}$$

Finally, you can compute the maximum tension at a sag of 35.23 feet on the catenary cable by using the equation $T = 893\,\text{k} + 1\,\text{klf} \cdot 35.23\,\text{ft} = 928.23\,\text{k}$.

That's slightly higher than the tension of the parabolic case. This discrepancy makes sense because you actually have slightly more total load acting on the catenary system due to the fact that the arc length of the cable is always longer than its horizontal dimension.

The sag on the parabolic system was given as 35.0 feet, so the same geometry conditions applied to the catenary cable only increased the sag by 0.23 feet.

For this example, the cable is much longer in comparison to the amount of sag. For problems with very small sag values, the arc length of the cable is very nearly the same as the horizontal distance between the supports. As the sag decreases, the total load on the catenary cable structure becomes more similar to the horizontal load of the parabolic cables. The difference in catenary behavior comes when the sag of the system becomes much larger.

Taking a Shortcut: The Beam Analogy for Flexible Cables

When working with cable systems, the first major piece of information you need to determine is the horizontal component of the cable force which occurs at the location of maximum sag. The *beam analogy for flexible cables* is a simplified technique for determining the horizontal tension component.

Before you use the analogy, though, keep the following basic assumptions in mind:

- ✔ All loads must be vertical, but they can be either concentrated or uniformly loaded.

- ✔ Any distributed load must be acting on a horizontal length, so this method doesn't work for catenary cables.

The beam analogy, as the name implies, requires that you create an analogous beam with the same loads as the cable structure. To implement this process, you start by following three simple steps:

1. **Create a horizontal beam having the same length as the cable system between support reactions.**

2. **Apply all external loads from the cable structures on the beam at the same horizontal location and show the vertical reactions of the beam.**

   You can't compute the horizontal reactions yet, but that's okay.

3. **Draw the moment diagram for the loadings of the beam.**

From the moment diagram (see Chapter 20), you can then make use of the following relationship:

$$(T_o)(y_{max}) = M_{BEAM}$$

The maximum sag occurs at the location of the horizontal tension component T_o. The moment M_{BEAM} isn't necessarily the maximum moment on the moment diagram; rather, it corresponds to the moment at the point on the moment diagram where T_o is to be calculated.

Consider the cable structure shown in the real system portion of Figure 22-8, with sag as shown. Two 30-kip forces are applied at 15-foot increments, one each at both Point B and Point C.

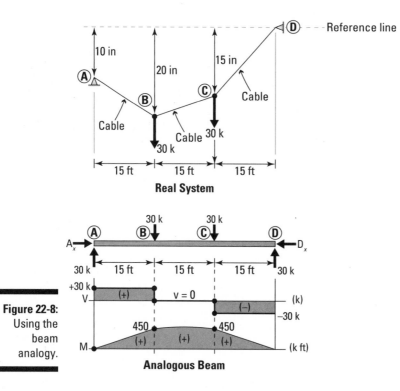

Figure 22-8:
Using the beam analogy.

Now you're ready to create the moment diagram for the analogous beam as shown in the analogous beam portion of Figure 22-8. The beam has two support reactions A_x and A_y acting at Point A, and two support reactions D_x and D_y acting at Point D, and an applied 30-kip point loads acting at both Point B and Point C.

From equilibrium, you then compute both of the vertical reactions (you need both to draw the moment diagram). I choose to sum moments at Point A in order to find the vertical reaction at Point D, D_y first.

$$\circlearrowleft + \sum |\overrightarrow{M_A}| = 0 \rightarrow -30 \text{ k}(15 \text{ ft}) - 30 \text{ k}(30 \text{ ft}) + D_y(45 \text{ ft}) = 0 \rightarrow D_y = 30 \text{ k}$$

You can use symmetry to discover that $A_y = D_y = 30$ k.

Finally, the shear and moment diagrams are drawn, and you see that at the location of maximum sag (Point B), the applied moment M_{BEAM} is 450 kip-feet. After you have this moment computed, you can then calculate the internal tension T_o in the cable by using the given sag at the location (or 20 inches in this example).

$$(T_o)(20 \text{ in}) = 450 \text{ k} \cdot \text{ft}\left(\frac{12 \text{ in}}{\text{ft}}\right) \rightarrow T_o = 270 \text{ k}$$

After you compute the tension T_o, you can then proceed on a segment-by-segment base and determine the magnitude of the tension in each segment (or at all locations).

If you know the cable tension and want to find the maximum sag, you can also compute that from the previous equation by entering the appropriate value of T_o and then solving for the unknown maximum sag value!

Chapter 23

Those Darn Dam Problems: Submerged Surfaces

. . . .

In This Chapter
▶ Defining fluid pressure parameters

▶ Explaining hydrostatic pressure

▶ Performing submerged surface calculations

▶ Incorporating openings and gates in submerged surface analysis

. . . .

*B*ecause most civilizations use submerged surfaces such as dams to control flooding and manage their water resources, engineers must understand pressure changes and design their equipment to perform under extreme situations.

When you dive into the deep end of a swimming pool, you can feel the pressure pushing all around, and the deeper you swim (or sink, if you're not a strong swimmer), the stronger the pressure you feel. At large depths (such as the bottom of the ocean), these fluid pressures can be downright deadly, which is why professional deep-sea divers must use specialized equipment to survive.

In this chapter, I show you some of the basic calculations that you can perform on a submerged surface. I explain the types of forces that are created by fluid pressures and how to calculate their quantities. I also show you how to apply fluid pressures to your free-body diagrams (F.B.D.s). Finally, I explain how to calculate partial fluid pressures on gates and openings.

Feeling the Pressure: Understanding Fluid Pressure

For the purposes of this text, I deal only with *incompressible fluids* (or fluids that don't change volume) such as water. The study of incompressible fluids and their pressures can be broken into two categories: dynamic fluids and static fluids.

✔ **Dynamic fluids:** *Dynamic fluids* include all fluids in motion or subject to a pressure. The flow of water within the water lines inside your home and the flow of gasoline from the pump to your car are examples of a dynamic fluid.

✔ **Static fluids:** *Static fluids* include fluids at rest or fluids subjected to non-pressurized flow. Examples of static fluids include a lake or reservoir and the milk in your morning cereal bowl.

In this text, I deal exclusively with static fluids. The forces from static fluids are classified into two categories: forces from hydrostatic pressure and forces from fluid self weight.

• **Hydrostatic pressure:** *Hydrostatic pressure* is the pressure associated with the depth of the fluid below the fluid surface.

• **Fluid self weight:** *Self weight* is gravitational effects acting on the mass particles of the fluid.

In Chapter 9, I define a relationship relating the specific weight (γ) of a material to its density (ρ): $\gamma = \rho g$, where g is the gravitational acceleration constant.

The material properties of most fluids are dependent on the temperature of the fluid, but for the purposes of this text, I take them as the following:

✔ **SI units:** ρ = 1,000 kilograms per cubic meter and γ = 9,810 Newton per cubic meter

✔ **U.S. customary units:** γ = 62.4 pounds per cubic foot. (The U.S. customary units for ρ are really ugly, so I just use specific weight in U.S. customary units.)

Dealing with hydrostatic pressure

Hydrostatic pressure is the pressure of a fluid associated with its depth and is a function of the type of fluid, the gravitational constant, and the depth of the fluid below the fluid surface. Figure 23-1 shows a typical hydrostatic pressure distribution for a fluid.

Hydrostatic pressure is a linear distribution that starts at a value of zero at the fluid's surface and increases linearly with vertical depth. The relationship between pressure p and depth z is given by $p = \rho g z = \gamma z$, where ρ is the density of the fluid, g is the gravitational constant, γ is the specific weight of the fluid, and z is the depth below the surface of the fluid.

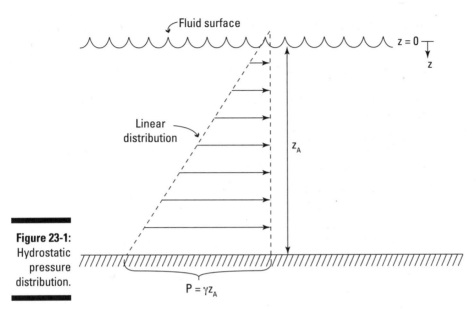

Figure 23-1:
Hydrostatic
pressure
distribution.

So at an arbitrary depth of z_A, the pressure is γz_a. At a depth of 10 feet, the hydrostatic pressure is $p_{10} = (62.4 \text{ pcf}) \cdot (10 \text{ feet}) = 624$ psf.

The *resultant* (or total force) of this entire distribution occurs at a distance of (depth/3) above the bottom. This distance is also the *centroid* (geometric center — see Chapter 11) of a triangular distribution. For more information on resultants of distributions, flip to Chapter 10.

Recognizing why zero pressure isn't exactly zero pressure

In pressure calculations, you have two types of pressure readings: absolute pressure and gauge pressure.

- ✔ *Absolute pressure* is the pressure exerted on an object, including a body of water, by the air and atmosphere above it. However, over small areas, this pressure is usually fairly constant and is taken as an approximate value of 14.68 pounds per square inch or 101.325 kilopascals.

- ✔ In submerged fluid calculations, you usually make another type of pressure calculation known as *gauge pressure,* which is a measure of the pressure above atmospheric pressure. The gauge pressure at the surface of a fluid is taken to be zero (because the depth is zero at the surface) and increases with depth.

Working with a unit width

In Chapter 10, I explain that all linear distributed loads have an intensity that is measured in units of force per length. Notice, however, the hydrostatic pressure calculation earlier in the chapter produced a pressure that was measured in pounds per square foot (which is a force per area unit), which are not the correct units for a linear distributed load.

To work around this units issue, you need to include a width dimension. Submerged surface problems (especially dams and wall structures) are often very long or have an unspecified total length, so you typically assume the forces you're calculating are acting along a *unit width,* which is equal to either 1 foot or 1 meter (depending on your system of units). Multiplying the previous pounds per square foot units by a unit width of 1 foot produces units of pounds per linear foot, which satisfies the units for a distributed load.

Determining effects from the self weight of water

The self weight from a fluid is the weight of a region of fluid acting directly above the object of interest and is the second source of load from fluids. Imagine collecting all the water above an object and storing it in a bucket. If you place that filled bucket on a scale, clearly it has weight which you must apply to the object of interest. Consider the dam shown in Figure 23-2 subjected to a water depth of 10 feet and a horizontal width on the face of the dam of 6 feet.

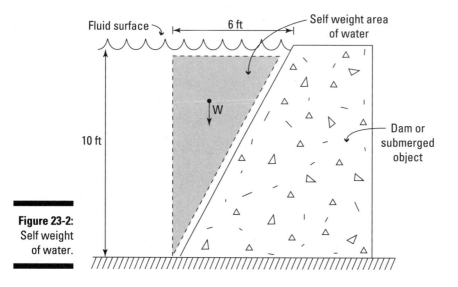

Figure 23-2:
Self weight
of water.

If you want to determine the weight of water acting on the dam, you include the volume of water multiplied by the specific weight of the fluid:

W = (Specific weight of fluid) · (Volume of fluid).

If you don't know the width of the dam, you'll need to convert the volume calculation to an area calculation multiplied by a unit width, as shown in the following equation:

W = (Specific weight of fluid) · (Two-dimensional area) · (unit width)

If you plug in the numbers for Figure 23-2, you get

W_w = (62.4 pcf) · (0.5 · (10 ft) · (6 ft)) · (1 ft) = 1,872 lb per unit width

The resultant force of this self weight occurs at the center of mass (also the centroid) of the area of the fluid, or in this case at ⅔(6 feet) = 4 feet from the front of the dam.

When you calculate the resultant of the water's self weight, you must also find the centroid of the water's area to know where the resultant is acting.

Making Calculations under (Fluid) Pressure

As with any statics problem, the first step is always to draw the correct free-body diagram, and submerged surface problems are no different. Check out the dam in Figure 23-3a.

Figure 23-3: Calculating fluid pressure on a dam.

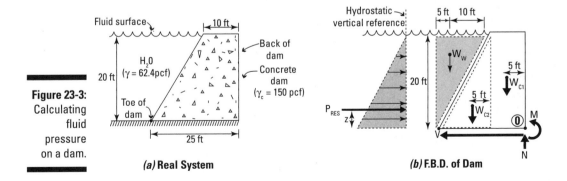

(a) Real System

(b) F.B.D. of Dam

To determine the loads acting on the dam's foundation, you must draw an F.B.D. — such as the one in Figure 23-3b — showing all forces acting on the dam, and the assumed *support reactions* (restraints), which are the foundation forces, consisting of a moment, shear, and normal force. The external loads appearing on this structure include the water forces from the linear hydrostatic pressure distribution, the self weight of the water above the dam, and the self weight of the concrete dam itself.

Drawing the fluid F.B.D.

When you're drawing a free-body diagram for a submerged surface, you must remember that both hydrostatic fluid pressure and self weight from fluids can occur simultaneously on the same structure. However, when the face of the structure is vertical, there is no fluid self weight acting on it. After all, there's no fluid area vertically above a vertical face.

Consider the sloped concrete dam (γ_{CONC} = 150 pcf) with dimensions shown in Figure 23-3. The dam holds back water (γ_{H_20} = 62.4 pcf) 20 feet deep. A design engineer needs to determine the forces on the base of the dam in order to design the proper foundation, so in the coming sections, I illustrate the steps required to compute these design values.

Creating the hydrostatic pressure distribution

The hydrostatic pressure for the problem in Figure 23-3 is a linearly varying distribution with a pressure of zero at the water surface and increases to $20(\gamma_{H_20})$ at the base of the dam.

This hydrostatic pressure acts along the entire height of the dam. However, to help keep the representation of this distribution clear, I like to draw it slightly off the structure horizontally and align the pressures relative to a vertical reference line. When I make this move, I can clearly see the distribution as a triangular distribution, and then I know how to compute the distribution's resultant. This trick is actually legal because from a statics point of view, all of the forces of the distribution are remaining on their original lines of action.

The first calculation you need to perform determines the hydrostatic fluid pressure distribution. The pressure at the water surface, where z = 0, is p_0 = (62.4 pcf) · (0 ft) · (1 ft) = 0 plf.

In Figure 23-3, you know that the depth of the water is 20 feet, so you can calculate the pressure intensity at the bottom of the distribution as $p_{20} = (62.4 \text{ pcf}) \cdot (20 \text{ ft}) \cdot (1 \text{ ft}) = 1{,}248 \text{ plf}$.

You can then compute the resultant force P_{RES} of the linear pressure distribution (see the preceding section) by determining the area of the linear hydrostatic pressure diagram:

$$P_{RES} = \tfrac{1}{2}(\gamma z) \cdot (z) \cdot (1 \text{ ft}) = \tfrac{1}{2}(\gamma)(z^2)(1)$$

In this case, the pressure distribution is triangular (or linearly distributed), with a value of zero pounds per linear foot at the water surface and 1,248 pounds per linear foot at the bottom. The resultant hydrostatic pressure for a unit width for this example is thus:

$$P_{RES} = \tfrac{1}{2}(62.4 \text{ pcf})(20 \text{ ft})^2(1 \text{ ft}) = 12{,}480 \text{ lb}$$

This load is acting at the centroid of the pressure distribution, which is located above the bottom at a distance of:

$$z = \frac{1}{3}(20 \text{ ft}) = 6.67 \text{ ft}$$

Finding the dead weight of water and dams

Next, you calculate the weight of the water volume acting directly above the concrete dam. This weight is applied as a single concentrated point load at the center of mass of the water volume (see Chapter 11 for more information about centroids and centers of mass).

Determining the self weight of water

To determine the weight of the water above the dam in Figure 23-3, you can compute the area of the water in two dimensions and multiply by the specific weight of the fluid and then multiply by the unit width, using the following formula:

$$W_W = (62.4 \text{ pcf}) \cdot \tfrac{1}{2}(15 \text{ ft}) \cdot (20 \text{ ft}) \cdot (1 \text{ ft}) = 9{,}360 \text{ lb}$$

This calculation provides the weight of water per unit width for the water acting directly above the concrete dam.

Establishing the self weight of a concrete dam

You also must include the self weight of the dam on the free-body diagram. For convenience of calculation, you can break the dam into two regions, a rectangular region with weight W_{C1} and a triangular region with weight W_{C2}. The calculations (which follow) are very similar to the calculation for the self weight of water (which you can find in Chapter 9) except that the specific weight for normal-weight concrete is approximately 150 pounds per cubic foot. For the rectangular region of the dam:

$$W_{C1} = (150 \text{ pcf}) \cdot (10 \text{ ft}) \cdot (20 \text{ ft}) \cdot (1 \text{ ft}) = 30,000 \text{ lb}$$

This weight is acting at a distance of 5 feet from the back of the dam. For the triangular region:

$$W_{C2} = (150 \text{ pcf}) \cdot \frac{1}{2}(20 \text{ ft}) \cdot (15 \text{ ft}) \cdot (1 \text{ ft}) = 22,500 \text{ lb}$$

This weight is acting at a distance of 10 feet from the front toe of the dam.

Including base reactions for dam structures

The reactions for the base of the dam can be modeled as a fixed support and have a horizontal force and a normal force (acting vertically) to prevent translation and a moment to prevent rotation.

The horizontal component of the reaction in Figure 23-3 is a shear force V acting parallel to the base along the interface of the dam and the foundation. Though you draw this force as a single concentrated load, it's actually spread along the entire length of the dam.

The vertical component of this reaction is a normal force N acting perpendicular to the base of the dam. At this time, the *point of application* (point where the force is acting in space) of this normal force is unknown. But for reasons I explain in Chapter 24, I assume it to be acting at the bottom corner at the back of the dam, or Point O in Figure 23-3b.

Finally, because all of the forces are *eccentric to* (or not acting at) the back corner of the dam, an equivalent moment M must be present at that point in order to maintain equilibrium and to prevent the dam from overturning.

Applying equilibrium equations

After you determine the F.B.D. of the dam in Figure 23-3 and apply the forces and support reactions (including the forces from the water and the self weight of the concrete dam), the final step is to actually apply the equations of equilibrium.

Shear *V* along the base of the dam is computed to ensure horizontal translational equilibrium:

$$\pm \sum \left| \overrightarrow{F_x} \right| = 0 \rightarrow P_{RES} - V = 0 \rightarrow 12,480 \text{ lb} - V = 0 \rightarrow V = 12,480 \text{ lb}$$

The positive sign on this calculation illustrates that the assumed direction of the base shear of the dam was correct. Next, the normal force *N* on the dam is computed to verify vertical translational equilibrium:

$$+\uparrow \sum \left| \overrightarrow{F_y} \right| = 0 \rightarrow -W_W - W_{C1} - W_{C2} + N = 0$$
$$\Rightarrow -9,360 \text{ lb} - 30,000 \text{ lb} - 22,500 \text{ lb} + N = 0$$
$$\Rightarrow N = +61,860 \text{ lb}$$

Finally, you can compute the *overturning moment* (or the moment that must be resisted to prevent the dam from toppling) *M,* by summing moments about the tipping point, or Point O, in Figure 23-3. You find the *tipping point* by examining the dam to determine which point the dam will rotate about should it start to fall over.

$$\curvearrowleft + \sum \left| \overline{M} \right| = 0 \rightarrow -P_{RES} \left(6.67 \text{ ft} \right) + W_W \left(20 \text{ ft} \right) + W_{C1} \left(5 \text{ ft} \right) + W_{C2} \left(15 \text{ ft} \right) + M = 0$$
$$\Rightarrow M = -591,000 \text{ lb} \cdot \text{ft}$$

The negative sign indicates that the moment direction assumed is backwards on the free-body diagram. This discrepancy actually has significant meaning with respect to overturning of the dam. As the negative sign implies, in order to actually make the dam overturn about Point O, an additional clockwise moment would have to be applied to overcome the weight of the water and self weight of the concrete dam. This finding means that the dam is in a stable (or equilibrium) condition, because in reality that additional moment doesn't actually exist. Hence, the dam can't overturn. Good news for the folks downstream, no doubt!

Figuring Partial Pressures on Openings and Gates

Dams and hydroelectric installations also employ systems of gates and valves to use forces from flowing water to spin turbines and provide electric power. These gates and openings allow water to flow through to mechanical machinery inside of the structure.

Unless the gate or opening is located at the water surface, the pressure distribution on the opening is no longer triangular, although it does remain linear. Instead, the pressure becomes a trapezoidal shape, with a unique pressure at the top of the opening and another at the bottom of the opening. Consider the *pivot gate* (a mechanical gate that is hinged or pivoted on one end) shown in Figure 23-4a.

The F.B.D. for the pivot gate contains the reaction forces B_x and B_y at the pivot and the normal contact force of the gate on the ground C_y. The external forces acting on the gate include the weight of the water volume, which has been broken into a rectangular and a triangular portion (W_{W1} and W_{W2}) respectively, and the trapezoidal hydrostatic distribution acting directly on the gate or opening.

This trapezoidal region has a pressure of γz_B per unit width at the top elevation of the gate and a pressure of γz_C per unit width at the bottom elevation. The following calculations give you the pressures for the trapezoidal region in Figure 23-4b:

$$\gamma z_B = (9,810 \text{ N/m}^3) \cdot (5 \text{ m}) \cdot (1 \text{ m}) = 49,050 \text{ N/m} = 49.1 \text{ kN/m}$$

$$\gamma z_C = (9,810 \text{ N/m}^3) \cdot (9 \text{ m}) \cdot (1 \text{ m}) = 88,290 \text{ N/m} = 88.3 \text{ kN/m}$$

To find the resultant, you can calculate the area of this trapezoidal region directly and then perform a separate calculation to determine the centroid of the single force resultant. However, a simpler method is to break this trapezoidal region into a simple rectangular and triangular region and find the resultant and location of each region individually (keeping the math calculations a lot simpler!).

In the example of Figure 23-4, P_{RES1} represents the resultant of the rectangular subregion and P_{RES2} represents the resultant of the triangular subregion. You can calculate P_{RES1} per unit width from the uniform distribution, which has an intensity of γz_B and height of $(z_C - z_B)$.

$$P_{RES1} = (49.1 \text{ kN/m}) \cdot (9 \text{ m} - 5 \text{ m}) = 196.4 \text{ kN}$$

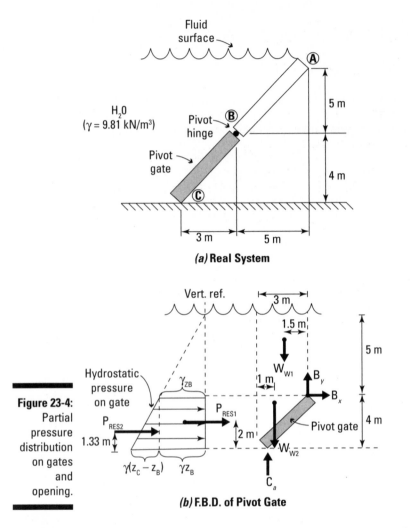

(a) Real System

(b) F.B.D. of Pivot Gate

Figure 23-4:
Partial
pressure
distribution
on gates
and
opening.

Because this distribution is rectangular, you also know that its resultant location is at a height of 2 meters (or one half of the total) from the base of the distribution.

Similarly, you can calculate P_{RES2} per unit width from the triangular region, which has a maximum intensity of $\gamma(z_C - z_B)$, and a height of $(z_C - z_B)$.

$$P_{RES2} = \tfrac{1}{2}(9.81 \text{ kN/m}^3) \cdot (9 \text{ m} - 5 \text{ m}) \cdot (9 \text{ m} - 5 \text{ m}) \cdot (1 \text{ m}) = 78.5 \text{ kN/m}$$

You know that the centroid of the triangular region will be at a distance of one third of the height of the opening above the base of the opening.

The self weight of the water acting on the gate extends from Point C to Point B vertically up to the water surface in Figure 23-4a. In this example, the volume of water is a trapezoidal area with a unit width. Just as with the trapezoidal hydrostatic pressure I discuss earlier in this section, you can also easily break this trapezoidal volume into a rectangular region and triangular region.

You calculate the weight of the water in the rectangular region per unit width, W_{W1}, from $W_{W1} = (9.81 \text{ kN/m}) \cdot (3 \text{ m}) \cdot (5 \text{ m}) \cdot (1 \text{ m}) = 147.2 \text{ kN}$ and is located at a distance of 1.5 meters horizontally from Point B (or half of the 3-meter horizontal dimension of the pivot gate). The weight of the water in the triangular region, W_{W2}, per unit width is calculated as

$$W_{W2} = \frac{1}{2}(9.81 \text{ kN/m}) \cdot (3 \text{ m}) \cdot (4 \text{ m}) \cdot (1 \text{ m}) = 58.9 \text{ kN}$$

which acts at a horizontal distance of 1 meter (to the right) from Point C.

At this point you have all the forces computed that are acting on the gate. You can now apply the equilibrium equations as I do in the preceding section to compute the vertical reaction at Point C, or the forces on the pin at Point B, depending on what information you want to determine.

Chapter 24

Incorporating Friction into Your Applications

When you think of friction, perhaps you think of the heat that's generated when you rub your hands together on a cold morning, or the scrapes and burns on your knees when you trip and fall over your dog on the way to the refrigerator in the middle of the night. Although these examples can be correctly called *friction,* in statics it has a slightly different meaning.

As you read through this chapter, you may realize that its contents are significantly different from the other chapters in Part VI of this text. In the other chapters of this part, I neglect the effects of friction on the problems in order to better explain the various techniques used to solve statics problems. However, in the real world, friction is an ever-present force that you must account for in all your calculations, so this chapter shows you how to do just that.

In this chapter, I describe the different types of friction and how to calculate their magnitudes, and then incorporate these values on a free-body diagram (F.B.D. — see Chapter 14). I also explore how the different objects in statics are affected by the various forms of friction.

Friction: It's More Than Just Heat!

In the old-time Western movies, a popular image shows a bartender sliding a beverage along the bar to a waiting customer at the other end. Somehow, magically, the bartender is able to hurl it down the bar and somehow make it stop exactly in front of his waiting patron. According to Newton's first law (which I discuss in Chapter 16), if the mug is moving down the bar, it continues in the same direction until an outside force acts on it. In this movie, that outside force is *friction* and it's what the bartender is counting on to save his beverages.

You can often blame friction for all sorts of funky behaviors. Stationary objects that refuse to move and tall objects that topple rather than slide are all victims of friction in one way or another.

Factors affecting friction

Friction is caused by several factors that occur on the surface of all materials. One factor that affects friction is the combination of microscopic imperfections on the interfaces of all materials that rub and interlock with each other as one object slides past another. The rougher the surface, the more frictional resistance that can occur. Another factor that is present in friction is the *adhesion* (or stickiness) between materials. Adhesion is affected not only by the materials in contact but also the presence of lubricants (or lack thereof) on the contact surface.

Other factors that can affect frictional resistance include applied normal force, characteristics of the contact surface, and length/area of the contact surface.

- ✔ **Applied normal force:** An *applied normal force* is the force acting perpendicular to the contact surface.

- ✔ **Characteristics of the contact surface:** The frictional characteristics of a surface are measured by a numeric constant called the *coefficient of friction* (which is simply a ratio of force required to move an object to the contact forces between the friction surfaces).

- ✔ **Length/area of contact surface:** The more contact length between two objects, the more interlocking and adhesion that can be developed.

Friction caused by two objects rubbing past each other is also known as *dry friction* (sometimes referred to as the *Coulomb friction force*) and is given by the inequality $F \leq \mu_s N = F_{MAX}$, where *F* is the resisting friction force and F_{MAX} is the force that must be overcome before an object can move. μ_s is the coefficient of static friction. *N* is the contact force (or the normal force) perpendicular to the interface between the two objects.

Types of friction

In mechanics, you actually run into two types of friction:

- ✔ **Static friction:** *Static friction* is the friction that resists applied forces before an object begins to move.

- ✔ **Dynamic friction:** *Dynamic friction* (also known as *kinetic friction*) is the friction that resists applied forces after the object has already started moving.

Because all the problems addressed in statics are typically not moving, in this book I focus exclusively on problems involving static friction. Table 24-1 shows approximate values for several common coefficients of static friction assuming a dry and smooth surface condition for each material.

Table 24-1	Common Approximate Coefficients of Static Friction
Material	*Coefficient*
Steel on steel	0.8
Glass on glass	0.9
Teflon on Teflon	0.05
Aluminum on steel	0.6

You can greatly reduce the coefficient of static friction between surfaces by lubricating them with oil, grease, or water.

A Sense of Impending . . . Motion? Calculating Sense

When you calculate the size of the static friction force F_{MAX} (which happens to also be its magnitude) by using the coefficients of static friction and you determine that the force's location (or *point of application*) must be along the interface between the objects, you've fulfilled two of three requirements (see Chapter 4) for defining a force vector. But what about the *sense* (or direction) of the friction force vector?

Anytime you try to push a heavy object across the floor and it doesn't budge initially, you might move to another side and push from another direction. Regardless of which side of the object you push from, the fact that it doesn't initially move illustrates that the friction force is always fighting against you, or more specifically against the direction of intended motion, also called *impending motion.* This concept is the key to working with friction problems.

Impending motion refers to the direction that the object wants to move (or rotate) after it has overcome the resisting friction forces. Consider the block shown in Figure 24-1, which is subjected to a single horizontal load. From Newton's laws, you know that an object wants to move in the direction of a force applied to it.

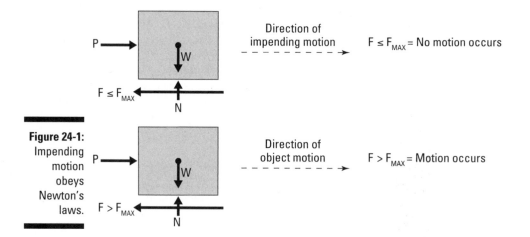

Figure 24-1:
Impending
motion
obeys
Newton's
laws.

After you've identified the direction of impending motion on an object, you're then ready to begin constructing the appropriate free-body diagrams and then you can start writing the equations of equilibrium. The following sections give you the lowdown on completing these tasks.

An F.B.D. of a problem including friction has all of the same information as a problem without friction. The major difference now is that you are also including the friction forces on the same diagram.

Establishing equilibrium when friction is present

Consider the block shown in Figure 24-2, which is subjected to a horizontal force P applied to the edge of a 1,000-pound crate. The coefficient of static friction is given as $\mu_s = 0.3$. The F.B.D. is shown.

Figure 24-2:
Creating a drawing and F.B.D. of a basic sliding object.

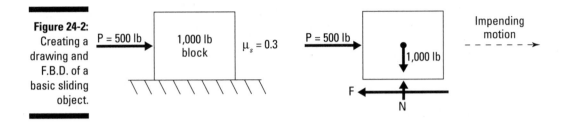

By looking at the applied force, logic tells you that the impending motion is in the direction of the applied load, or to the right. Because the object isn't currently moving (after all, the motion is impending, right?), you can apply the equations of static equilibrium:

$$+\uparrow \sum \left| \vec{F}_y \right| = 0 \rightarrow +N - W = 0 \rightarrow N = W \rightarrow N = 1{,}000 \text{ lb}$$

The first equation gives you the value of the normal force on the contact interface, $N = 1{,}000$ pounds. You use this normal reaction force to determine the friction limit F_{MAX} in the following section. You want to apply the other equilibrium equation (in the x-direction) to find the actual friction force acting on the interface due to the applied force P:

$$\xrightarrow{+} \sum \left| \vec{F}_x \right| = 0 \rightarrow P - F = 0 \rightarrow F = P = 500 \text{ lb}$$

This equilibrium equation tell you that if the applied load is 500 pounds, the resisting friction force must also be 500 pounds to ensure equilibrium.

Finding the friction limit F_{MAX}

To determine if the object moves due to the 500 pound load, you need to examine whether that load is more than the friction limit F_{MAX}:

$$F_{MAX} = \mu_s N$$

REMEMBER

If the friction force F at the interface is less than the friction limit F_{MAX} (which is the limiting force before motion occurs), the object doesn't move. If the friction force exceeds the friction limit, there is enough force on the object to overcome friction and cause it to start moving in the direction of impending motion.

Plugging the numbers for Figure 24-2 earlier in the chapter into the formula, you get $F_{MAX} = (0.3)(1{,}000 \text{ lb}) = 300 \text{ lb}$.

This result indicates that in order for the object to slide, the friction force F must be greater than the friction limit $F_{MAX} = 300$ pounds. Thus:

$$F = 500 \text{ lb} > F_{MAX} = 300 \text{ lb}$$

Thus, in this example, you've determined that you have sufficient applied force to overcome the friction limit, and the crate therefore moves.

Solving Friction Problems by Using Logic and Equations Together

Sometimes, contact points on an object can move in different directions. An example of this situation is shown in Figure 24-3a, which presents a 150-Newton plank 4 meters long resting on the ground at Point A and against the wall at Point B. The coefficient of static friction at Points A and B is $\mu = 0.3$.

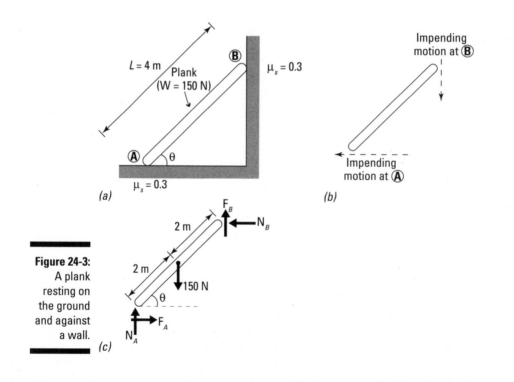

Figure 24-3: A plank resting on the ground and against a wall.

To draw the complete F.B.D. for this plank, you must account for friction and normal forces at every contact point.

- ✔ The end of the plank at Point A wants to move to the left, and the top of the plank at Point B wants to slide down the wall. Thus, F_A, the friction at Point A, must be to the right to oppose the impending motion at Point A (see Figure 24-3b).

- ✔ At the same time, Point B wants to move downward due to the weight of the plank, and F_B, the friction force at Point B, must be acting upward.

- ✔ In addition to the friction forces, a normal force N_A occurs at Point A, and another, N_B, occurs at Point B.

In total, you have four unknowns acting on the F.B.D. of this plank. As it's currently drawn, this F.B.D. (see Figure 24-3c) is statically indeterminate to the first degree, meaning that there's one more unknown than there are available equilibrium equations.

For a two-dimensional equilibrium problem, you have only three equilibrium equations to work with. Fortunately, by taking advantage of the relationship between the friction limit and the normal contact force, you can both simplify the F.B.D. and provide extra equations in addition to equilibrium equations. I show you how to do this in the following section.

Working with friction angles

At every point where a friction force is applied, you always find a normal force applied at the same location. And for an object to move, the friction force must be greater than the friction limit: $F > F_{MAX} = \mu_s N$.

The friction limit is also a function of the normal force at that same location.

Together, the friction limit and the normal force form rectangular components of a single resultant. You can then calculate the magnitude of the resultant of these two forces as

$$R = \sqrt{(F_{MAX})^2 + (N)^2} = \sqrt{(\mu_s N)^2 + (N)^2} = N\sqrt{(\mu_s)^2 + 1}$$

which means that you can compute the resultant by knowing only the normal force and the coefficient of static friction (which is already a known quantity), as shown in Figure 24-4a. Employing the head-to-tail technique I describe in Chapter 6, you can calculate the magnitude R and the direction of the *friction angle*, ϕ, of a resultant of those two forces as shown in Figure 24-4b. All you have to do is apply a little bit of basic trigonometry.

$$\tan(\phi) = \frac{\text{parallel component}}{\text{perpendicular component}} = \frac{\mu_s N}{N} \rightarrow \phi = \tan^{-1}(\mu_s)$$

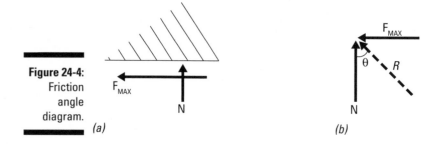

Figure 24-4:
Friction
angle
diagram.

Thus, the angle that the friction resultant makes with respect to the normal force is also a function of the static friction coefficient, or is a constant, because the normal forces in the calculation cancel each other.

Combining friction and normal forces into a single resultant

By creating a new single resultant for the friction limit and the normal force, you can eliminate having to work with two unknown forces. Instead, you can replace this system with a single unknown (in terms of the normal force) acting at a known angle, which you can compute from the coefficient of static friction.

For the earlier plank example of Figure 24-3, you can compute the friction angle as $\phi = \tan^{-1}(\mu) = \tan^{-1}(0.3) = 16.7°$.

The combined resultant at each contact point (because μ_s is the same at both points) is given as the following:

$$R = N\sqrt{(\mu_s)^2 + 1} = N\sqrt{(0.3)^2 + 1} = N \cdot 1.04$$

Thus, you can replace the friction and normal pairs at both Point A and Point B with a single resultant force at a new orientation. I've gone ahead and done this for you in Figure 24-5.

By replacing the normal and friction components with a single resultant R at a new angle ϕ at each contact point, you're now left with two unknowns, N_A and N_B, on the free-body diagram. Because you have three equations to work with, you're now able to solve for these components.

$$+\uparrow\sum\left|\vec{F_y}\right| = 0 \rightarrow 1.04\,N_A\cos(16.7°) + 1.04\,N_B\sin(16.7°) - 150 = 0$$

$$\underset{\rightarrow}{+}\sum\left|\vec{F_x}\right| = 0 \rightarrow 1.04\,N_A\sin(16.7°) - 1.04\,N_B\cos(16.7°) = 0$$

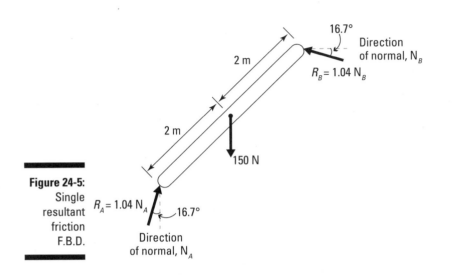

Figure 24-5:
Single
resultant
friction
F.B.D.

You can then solve these two equations simultaneously to give you the normal forces N_A and N_B that are required to keep the 150-Newton plank in equilibrium. The results: N_A = 138.2 Newton, and N_B = 41.4 Newton.

To determine whether the plank starts to slide, you then must calculate the friction limit for each contact point (flip to "Finding the friction limit F_{MAX}" earlier in the chapter for more on this calculation).

$$F_{A\,MAX} = \mu_s N_A = 0.3(138.2\ N) = 41.5\ N$$
$$F_{B\,MAX} = \mu_s N_B = 0.3(41.4\ N) = 12.4\ N$$

From this F.B.D., you can also calculate the actual friction forces at Points A and B by calculating the components of the resultant force and its orientation angle ϕ:

$$F_A = 1.04\ N_A \sin(16.7°) = 1.04(138.2\ N)\sin(16.7°) = 41.3\ N$$
$$F_B = 1.04\ N_B \sin(16.7°) = 1.04(41.4\ N)\sin(16.7°) = 41.2\ N$$

Because you know the actual friction force at each point, as well as the friction limit, you can start to draw some conclusions:

✔ Because F_A = 41.3 N ≤ $F_{A\,MAX}$ = 41.5 N, Point A can't move.
✔ Because F_B = 41.2 N > $F_{B\,MAX}$ = 12.4 N, Point B is able to move.

In order for the plank to move, both Point A and Point B must be able to move. After all, unless the plank stretches, one point can't move without the other also moving. In the Figure 24-3 example, based on the calculations, Point A can't move because its friction force isn't sufficient to overcome the friction limit (though it's actually very close). Even though Point B can move, you must remember that both points must move for the entire plank to move. Your final conclusion would thus be that the plank is unable to slide.

Timber! Exploring Tipping

Say you're back in your kitchen, trying to move that refrigerator yet again. First you push horizontally at a low point on the refrigerator (as shown in Figure 24-6a), and it begins to slide. Great! You push for a while and then take a breather. When you return to the task, you decide to push on the upper corner. But instead of sliding, what happens? Chances are, if it's a full-height refrigerator and it's empty, the refrigerator starts to lean and then fall over without ever sliding. This phenomenon of falling over before sliding is called *tipping*.

Uncovering the tipping point and normal force

To quantify tipping, you need to locate the *tipping point*, which is usually at a corner or edge along the contact surface of the object. When tipping occurs, the contact surface disappears, with the tipping point remaining as the last point in physical contact with the original interface (see Figure 24-6b).

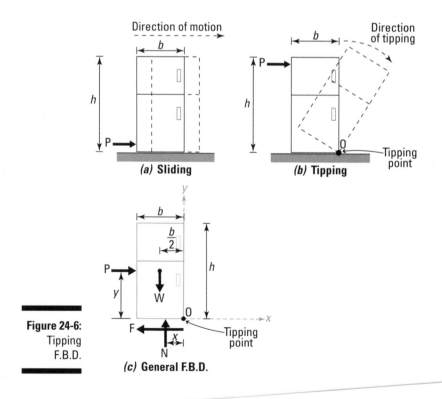

(a) **Sliding**

(b) **Tipping**

Figure 24-6:
Tipping
F.B.D.

(c) **General F.B.D.**

To begin analyzing tipping problems, you first need to construct an F.B.D. (surprise, surprise!). Referring to the free-body diagram in Figure 24-6c, you see that all of the old familiar F.B.D. information has been included: external forces, self weight, and support reactions (or contact forces and friction forces in this case). See Chapter 13 for more on the basic parts of an F.B.D.

However, a new parameter — location of the normal force x — has made an appearance. In most F.B.D.s, you assume the normal force is acting in line with the center of mass of the object and don't worry about the location of the normal. With tipping and friction problems, the location of the normal force matters. To start the analysis, you first apply the translational equilibrium equations from earlier in the chapter to determine the normal force N and the friction force F:

$$\xrightarrow{+}\sum\left|\overrightarrow{F_x}\right| = 0 \rightarrow P - F = 0 \rightarrow P = F$$

$$+\uparrow\sum\left|\overrightarrow{F_y}\right| = 0 \rightarrow N - W = 0 \rightarrow N = W$$

The applied force P and friction force F have the same magnitude but opposite directions, and they're separated by a distance y. Together, these two forces cause a *moment* (or a rotational effect that I discuss in Chapter 12) on the object, which causes the object to want to rotate in a clockwise direction about the tipping point. To maintain equilibrium, the normal force N shifts away from the center of the object in response, to create a balancing couple with the weight. The normal force N moves toward the tipping point (which causes x to decrease). If the force becomes too great, the distance x may actually become a negative value. When this situation happens, the normal must be located outside the boundaries of the object to balance the overturning moment from the applied load P, which is physically impossible. Your fridge is no longer stable and tips over.

Moving the normal force to prevent tipping

To continue your analysis of tipping, you first establish your Cartesian coordinate system (see Chapter 5) by placing the origin at the tipping point location. Then you write the rotational equilibrium equation for this F.B.D. about the tipping point (because tipping is a rotational behavior). For Figure 24-6, the equation is

$$\curvearrowleft+\sum\left|\overrightarrow{M_o}\right| = 0 \rightarrow -P \cdot y + W \cdot \left(\frac{b}{2}\right) - N \cdot x = 0$$

The terms of this moment equation indicate that each term is set by a physical parameter of the object or a known applied point load and doesn't vary, except for the position of the normal force with respect to the tipping point x. As the applied load P increases, the x parameter for the normal force N changes in order to maintain equilibrium. This position is what helps you determine whether an object will tip.

If you apply a force, say $P = 100$ pounds, at the base of the object (or $y = 0$), tipping doesn't occur. Say the fridge in Figure 24-6 has a base of 12 inches, is 60 inches high, and weighs 150 pounds. Substituting these values into the moment equation, you can then solve for x:

$$-100 \text{ lb} \cdot (0 \text{ in}) + 150 \text{ lb} \cdot \left(\frac{12 \text{ in}}{2}\right) - 150 \text{ lb} \cdot (x) = 0 \rightarrow x = +6 \text{ in}$$

Because x is positive, you know your assumption of the location of the normal force was correct. Now watch what happens if you move the same load to the top of the refrigerator, at $y = h = 60$ inches.

$$-100 \text{ lb} \cdot (60 \text{ in}) + 150 \text{ lb} \cdot \left(\frac{12 \text{ in}}{2}\right) - 150 \text{ lb} \cdot (x) = 0 \rightarrow x = -34 \text{ in}$$

Thus, by simply moving the force, you've also moved the location of the normal force; in this case, you've moved it to the right of the tipping point, or outside the physical boundary of the object, so the fridge tips over.

Note that the sign of x doesn't actually correspond to a Cartesian coordinate value. In this particular example, a negative x actually corresponds to a positive Cartesian position, so be careful!

Establishing which friction phenomenon controls, sliding or tipping

As the preceding section indicates, if you push horizontally at a low point on a refrigerator, it's more likely to slide. And if you push with the same force at the top of the refrigerator, it's more likely to tip. But what if you place your refrigerator on a sloped floor and start to increase the angle of the ramp θ without applying a force at all. Which friction phenomenon occurs first?

Consider the 300-pound refrigerator acting on the ramp, as shown in Figure 24-7a.

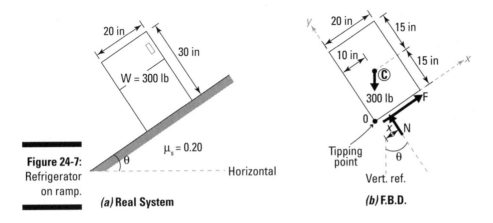

Figure 24-7:
Refrigerator
on ramp.

(a) **Real System**

(b) **F.B.D.**

To start your analysis, you must first write the equilibrium equations for the F.B.D. If you establish a coordinate system that is inclined with the ramp, with the origin at the presumed tipping point, you can create a simpler equation than if you had used regular horizontal and vertical components. Here's what the equations look like for Figure 24-7:

$$+\nearrow \sum \left| \vec{F_x} \right| = 0 \to F - (300 \text{ lb}) \cdot \sin(\theta) = 0 \to F = (300 \text{ lb}) \cdot \cos(\theta)$$

$$+\nwarrow \sum \left| \vec{F_y} \right| = 0 \to +N - (300 \text{ lb}) \cdot \cos(\theta) = 0 \to N = (300 \text{ lb}) \cdot \sin(\theta)$$

$$\curvearrowleft + \sum \left| \vec{M_o} \right| = 0 \to N \cdot (x) + (300 \text{ lb}) \cdot \sin(\theta) \cdot (15 \text{ in.}) - (300 \text{ lb}) \cdot \cos(\theta) \cdot (10 \text{ in}) = 0$$

Remember, these equilibrium equations are unique to each problem you solve. Just draw your F.B.D. (such as the one in the F.B.D. portion of Figure 24-7b) and then write the equilibrium equations based on your Cartesian axes.

Next, you list all the possible friction phenomenon that can occur. For this simple problem, you have two possibilities, which I discuss in the following sections.

Without knowing which occurs first, you must make a guess. Choose one of the two cases and then verify that the assumptions behind that case are correct. Friction requires a lot of guess-and-check type of calculations.

Case 1: Checking sliding before tipping

For sliding to occur first, you need to prove that the friction force at the bottom of the refrigerator exceeds the friction limit $\mu_s N$. For sliding to occur before tipping happens, the normal force is still located somewhere along the contact surface and the *x* dimension is positive. If sliding occurs, then

$$F > F_{MAX} \to (300 \text{ lb})\sin(\theta) > (0.3) \cdot (300 \text{ lb})\cos(\theta) \to \theta > 16.7°$$

Thus, at a ramp angle of 16.7 degrees, the refrigerator starts to slide. But now you must confirm your assumption about the location of the normal force x by checking the moment equation:

$$\circlearrowleft+\sum\left|\overrightarrow{M_o}\right|=(300\ \text{lb})\cdot\cos(16.7°)(x)+(300\ \text{lb})\cdot\sin(16.7°)\cdot(15\ \text{in})$$
$$-(300\ \text{lb})\cdot\cos(16.7°)\cdot(10\ \text{in})=0\Rightarrow x=+5.5\ \text{in}$$

The positive sign indicates that the normal force is assumed to be acting on the correct side of the tipping point, and more importantly, is still in contact with the interface.

Case 2: Checking tipping before sliding

For tipping to occur first, you assume that the location of the normal is at the tipping point (or $x = 0$) and then compute the corresponding friction force. Finally, to verify, you check that the friction force F is less than the friction limit F_{MAX}.

To check tipping for Figure 24-7, you start with the moment equation and assume $x = 0$ (that the normal is located at the tipping point).

$$N\cdot(0\ \text{in})+(300\ \text{lb})\cdot\sin(\theta)\cdot(15\ \text{in})-(300\ \text{lb})\cdot\cos(\theta)\cdot(10\ \text{in})=0$$
$$\Rightarrow\frac{\sin(\theta)}{\cos(\theta)}=\tan(\theta)=\frac{10\ \text{in}}{15\ \text{in}}\rightarrow\theta=33.8°$$
$$F=(300\ \text{lb})\cdot\sin(33.8)=167\ \text{lb}$$
$$F_{MAX}=(0.3)\cdot(300\ \text{lb})=74.8\ \text{lb}$$

If you compare this angle with the result from the preceding section, you see that you need a steeper angle to cause tipping than you do to cause sliding.

If you haven't already calculated the angle for sliding to occur (perhaps you chose to check tipping before sliding), you can check it now by determining the friction force F and comparing it to F_{MAX}.

Thus, $F > F_{MAX}$ at the angle required to cause tipping, which implies that sliding would have already occurred. The calculations in the preceding section verify this result.

Examining More Common Friction Applications

You encounter a wide variety of other types of friction problems in statics, but the most frequently encountered problems typically involve wedges or belt/pulley friction.

Wedging in on the action

Wedges are small mechanical devices that transmit (and usually increase) an applied force in another direction. Usually, wedges are very long in comparison to their thickness, which often eliminates tipping as a concern. (Flip to "Timber! Exploring Tipping" earlier in the chapter for more on this concept.) Consider Figure 24-8, which shows a force *P* being applied to Block A. As force *P* increases, Block A moves to the right. Because of the sloped interface between Blocks A and B, Block B moves upward as Block A moves to the right.

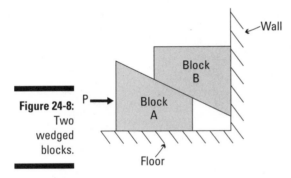

Figure 24-8:
Two
wedged
blocks.

To analyze problems involving wedges, you need to draw several F.B.D.s; you don't have to draw them in the order I list here, but make sure you draw them all:

1. **Draw one F.B.D. for each individual wedge.**

   Normally, you'd draw the combined system F.B.D. in Step 2 first, but in this example, looking at the separate diagrams first is helpful. You can actually determine the directions of the friction forces by looking at the individual behaviors of each block. Make sure to include them in the same directions on the combined system in Step 2.

Starting with Block A, you know that when the force is applied, the impending motion of Block A is to move to the right. (I discuss impending motion earlier in the chapter.) Thus, the friction on both surfaces of the wedge must oppose the motion of direction. The friction force F_A is to the left and the interface friction force F_{AB} is upward and to the left. You also must include both the normal force from the ground (N_A) and the normal force from Block B sitting on top (N_{AB}). See Figure 24-9.

Normal forces always act perpendicular to their contact surfaces.

Similarly, you know that Block B wants to move upward, so the friction forces on it must oppose that impending motion. Thus, the friction at the wall (F_B), is acting downward, and the interface friction (F_{AB}) is acting down and to the right. (Notice that the force F_{AB} on Block B is in the opposite direction of the same force on Block A, which ensures equilibrium.)

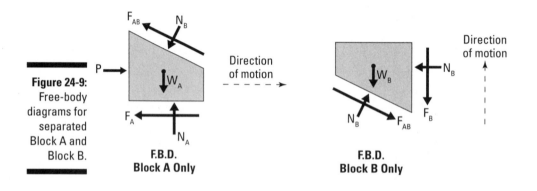

Figure 24-9: Free-body diagrams for separated Block A and Block B.

F.B.D. Block A Only

F.B.D. Block B Only

2. **Draw an F.B.D. of the combined systems.**

You also need to draw a combined diagram of Blocks A and B. Remember to include the normal and friction forces from the contact surfaces with the wall and floor on your F.B.D.; you can get those from the individual block diagrams in Step 1. See Figure 24-10.

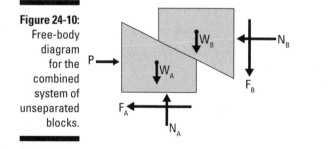

Figure 24-10: Free-body diagram for the combined system of unseparated blocks.

3. **Write equilibrium equations for each individual block and the combined F.B.D.**

   You sum forces in the *x*- and *y*-directions for each of the three diagrams in Steps 1 and 2 to develop relationships between unknown forces and the applied load *P*.

4. **Check the friction limits for each friction force.**

   Remember, in order for Block A to move, the friction at the floor and at the interface with Block B must be larger than the friction limits at each location. If Block A can move, you must then check that Block B can also move. If Block B doesn't move, Block A can't move.

Staying flexible with belts and pulleys

Another type of friction problem that you encounter involves pulleys and *belt/cable friction*. This friction is caused by the motion of a belt or cable relative to the surface of a pulley or drum assembly (see Figure 24-11). The direction of movement of the cable determines the direction of the friction force.

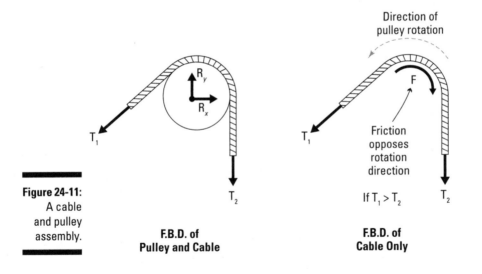

Figure 24-11: A cable and pulley assembly.

F.B.D. of Pulley and Cable

F.B.D. of Cable Only

However, you normally express the force on one side of the pulley in terms of the force on the other when examining belt friction problems.

For frictionless cases, the tension on both sides of a pulley or drum is assumed to be equal. If friction is present, that's no longer the case. The tension of the belts on either side of a pulley subjected to belt friction is given by the following equation:

$$T_1 = T_2 e^{\mu\beta}$$

where β is the *angle of wrap* of the belt around the pulley, expressed in radians. If the belt wraps 180 degrees, the angle of wrap is π or 3.14 radians. The larger the angle of wrap, the more force is necessary to overcome friction.

Part VII
The Part of Tens

In this part . . .

Here comes the fun stuff. In this part, I include two top-ten lists that give you a quick reference for how to tackle a basic statics problem and ten important guidelines to remember when taking a statics exam.

Chapter 25

Ten Steps to Solving Any Statics Problem

In This Chapter
▶ Including all the important parts in your drawing
▶ Making common assumptions

*Y*ou're walking down the street, confident in your newly honed statics skills, when you see a big mass of support reactions and applied loads unlike any you've ever seen run out of the building in front of you. It quickly turns, let's out an evil laugh, and rushes straight for you. Your first instinct is to turn and run (and who can blame you really, some statics problems can be especially nasty). Or you can sidestep the problem altogether and pretend to not notice, but that means letting an unsolved statics problem run rampant through the world. There are innocent bystanders watching, and you must ask yourself, "What do I do?"

Armed with paper, pencil, and calculator, you set to work to save all of humanity. Now, if only you could remember the vital aspects of statics problem solving. Luckily, this chapter provides the ten concepts you need to remember for taming that runaway problem.

Sketches Come First

Without a doubt, when in doubt, the first thing you must do is quickly make a sketch of the statics problem. You're not trying to pick it out of a police lineup; a sketch just gives you the best basic starting point for static analysis. When you first start sketching, don't worry about all of the little details and the moving internal parts — just make a quick sketch of the object as a whole and focus simply on how the object is attached to the world around it. For now, disregard any internal features such as internal hinges, pulleys, and machine parts.

Determine the Supports

The attachments of your monster to the world around it represent the *support reactions* (restraints). By determining the support reactions, you can actually reduce the scale of the free-body diagram (F.B.D.) to a more manageable size. (Head to Chapter 13 for more on supports and F.B.D basics.)

If your statics behemoth is on wheels or is sliding toward you, you're dealing with a support with only one contact force (such as a *roller support*). If a support isn't moving, you need to model that as either a pinned or fixed condition; a *fixed support* isn't rotating either, whereas a *pinned support* may be. If you don't know the type of support for sure, just assume it's fixed.

Don't Forget the Applied Loads and Self Weight

Look at the problem, and see what's causing it to move forward. Is a point load or distributed load acting on it? Does it have water pressure pushing on it? Is something causing an applied torque or tensile force attached to it? These possibilities are all important considerations in your solution. Additionally, don't forget to include any *self weight* (the force created by gravity's effects on the object's mass — refer to Chapter 9).

You need to use all of this information to construct a *free-body diagram* (or the detailed diagram that contains all of the loads and dimensions necessary for performing a static analysis).

After you determine the loads and support reactions (see the preceding section, you can make a basic free-body diagram of the entire system by using the concepts in Chapter 13.

Calculate As Many Unknown Support Reactions As You Can

After you've created a free-body diagram of a statics problem, your next step is to determine any unknown support reactions even though you haven't started looking at the internal features at all. After all, if you're going to cut a huge object loose, you want to have some idea of the size and direction of the forces that were holding it back in the first place.

When the object was restrained at the supports, it may have been struggling to break free, but it was in a balanced state. That means you can apply Newton's laws of motion, or your equations of equilibrium from Part V.

Enforcing the translational equilibrium equations is pretty straightforward. You simply add up the forces in a given direction and write the expressions. If you're lucky, you may be able find a reaction or two, or at the very least create a relationship between them.

The final equation that you want to write is the moment equation, and this one is where you have some control over a beastly problem. Depending on how you attack it, you can either make things a whole lot worse or solve for unknowns outright. In simple problems (such as *statically determinate* problems, which are problems that have sufficient information to be solved by just the basic equilibrium equations), you can usually sum moments at a pinned support and knock out two of the total unknown forces from the moment equation. However, you have to be more alert with *statically indeterminate* problems. If you have more than three unknown support reactions, you have to find a point on the lines of action of as many of these reactions (or an *instantaneous center*) as possible to choose as your summation point. (Check out Chapter 19 for more on this topic.)

For the super crazy problems, you may not be able to determine any of the support reactions ahead of time. In fact, some problems can't be solved at all by statics alone. For those problems, you need friends from other mechanics subjects to help! But don't give up hope just yet!

Guess It's a Frame or Machine

Before you can decide how to tackle a statics problem, you need to be able to identify it. Think of yourself as a medical doctor specializing in the treatment of bizarre statics problems. You have to first identify the underlying cause (identify the type of structure) before you can begin treatment (write and solve equations). To accomplish this task, follow this handy checklist:

- ✔ **Is this problem a truss?** Trusses are pretty simple to identify. Are all members of the system connected at the ends (or joints) only through internal hinges? Are all loads on the system concentrated forces and are they applied only at joints? If you answered *yes* to both questions, congratulations — you have a truss. You can use the principles of Chapter 19 to deal with this problem. If the answer to either of those questions is *no*, you need to ask yourself a few additional questions because you don't have a truss.

- ✔ **Is this problem a submerged surface problem?** This one is typically fairly obvious. If you have a fluid involved (whether it's water, oil, or whatever), you're dealing with a submerged surface and can refer to Chapter 23.

✓ **Is this problem a cable problem?** Another fairly obvious diagnosis, doctor. If only a rope or cable system is supporting a load, you're dealing with a cable system. You can handle those fellas by looking to Chapter 22.

If none of the checklist categories fits your problem, you probably have a system such as a beam or a frame and machine. You can actually solve for internal forces of both of these types of categories using the same principles. If you slice a member and reveal the internal forces, both of these problem types have three internal forces — axial, shear, and moments — at every cut location. Chapters 20 and 21 give you solution ideas for these problem types.

If you're not sure about the type of problem, always assume you have three internal forces at every cut location.

Get Out the Dynamite: Separating Pieces from the Problem for Internal Analysis

When you're ready to dissect the statics problem and look at what's happening internally (after all, the internal forces are usually the most important for design), you have a couple of different options. If you're dealing with a system that contains internal hinges, many of the major methods of analysis involve breaking a structure into smaller pieces. You can run down to the old ACME mine and grab a friendly barrel of dynamite, light the fuse (get a long one!), and run for cover. Of course, the end result is a lot of smaller pieces, and unless you like jigsaw puzzles, you have a few more free-body diagrams to draw with this tactic than may otherwise be necessary. Instead, consider a more surgical approach.

Instead of blowing the structure to smithereens (which is slight overkill for most statics problems), look for pieces of the structure that you can easily separate from the main system in a more controlled manner. Items such as mechanical attachments (blades, presses, pistons, and so on), cables, and pulleys are all prime candidates for extraction. These items are usually hinged at their connection points to allow them to rotate. Hinges prove to be very useful for removal of objects because you know the moment is always zero at these locations. If you know a location of zero moment, it will prove to be a useful place to separate the structure because at these locations, you no longer have an unknown internal moment to deal with in your moment equilibrium equations when you cut the structure. Apply the basic equilibrium equations to find the internal hinge forces, and you're well on your way to analyzing the structure.

Assume Directions of Internal Forces

Remember that to calculate internal forces, you first have to slice a few members, which allows you to draw additional free-body diagrams and gives you additional equilibrium equations to work with. So you follow the same basic steps for applying support reactions, applied loads, and self weight, but in addition you must include the internal forces from each and every location.

The problem is that without writing the equilibrium equations, you don't know the values of the exposed internal forces at the time that you're drawing the free-body diagrams. Most of the time you don't even know the direction of the internal forces either. So what do you do?

The first thing you should do is try to determine the type of member that you've cut. If you're dealing with a cable or a rope (see Chapter 22), you know that those two systems are both axial-only systems. Furthermore, you also know that cable and rope systems' internal forces are always in *tension* — the direction of the force is pulling on the object.

If the member you've cut is a truss member (Chapter 19), you know that the forces in that member are also axial-only. However, truss members may have either axial tension or axial compression loads. When you're drawing your free-body diagram, you don't know whether the member is in compression or tension until you start to write the equilibrium equations, so the common convention is to assume that the forces in the member are acting in tension when dealing with trusses. You then look at the equilibrium equations to confirm whether this assumption is true — if the numerical value is negative, the direction you assumed is incorrect.

Cutting most other members exposes an axial force, as well as a shear force and a moment. As with axial members, the common assumption is that the force is acting in tension. Typically, for shear and moment, you can use the positive sign convention for internal bending forces that I describe in Chapter 20. At this point, what's most important isn't the direction of these internal forces but rather that you have at least included them on the diagram. In the end, regardless of which direction you assume is positive, the signs of the numerical values from the equilibrium equations always confirm or refute your previous assumptions.

Be Consistent with Your Assumptions

Remain consistent in the directions you assume. If you always assume axial forces are tension, shear forces are positive, and moments are counterclockwise, when you solve the equations for the actual numerical values, a negative sign has the same meaning every time. If you vary your assumptions, you have to keep checking your diagram to verify what direction you assumed where and how the sign affects it.

Guess That Three (or Six) Equilibrium Equations Are Necessary

When you work toward finding internal forces and you have created additional free-body diagrams, the next step is to write the equations of equilibrium.

For two-dimensional problems, you have two translational and one rotational summation that you need to make (for three total equations). For three-dimensional problems, you have three translational and three rotation summation equations, or six total equations.

The more free-body diagrams you make, the more equations you have to work with.

If Friction Is Involved, Guess That the Object Slides

The most difficult statics monsters to deal with are those that involve friction. Many common problems neglect friction, but those that don't are more complex animals. Free-body diagrams of friction problems have extra unknown forces acting on them and may even become indeterminate.

As I explain in Chapter 24, tipping and sliding problems always have a friction force in the direction of the motion of the object at the boundary or *interface* of every contact surface. The normal contact force location is now at a variable location that's the key to determining whether the force causes the object to slide or to tip over. To get started, assume that the friction force at the interface is equal to the friction limit at that surface and then use that force to calculate the location of the normal contact force. If the contact force location keeps the contact force on the object, the object slides. If the contact force is outside the boundary of the object, the object tips, and your original sliding assumption is wrong. To correct this inaccuracy, place the normal force at the tipping point and resolve the problem as I show you in Chapter 24.

Chapter 26

Ten Tips for Surviving a Statics Exam

In This Chapter

▶ Remembering vital concepts

▶ Getting as much partial credit as possible

Although any science or math test can be a frustrating or intimidating experience, a statics test can reach a whole other level of intimidation. In this chapter, I present ten suggestions that help make your life a little easier during those stressful statics exams.

Find Problems You Know How to Solve

After you've received your exam (and you've taken a deep breath to help you gather your composure), flip through the test and quickly read each problem, highlight what you're being asked to find, and then choose a problem that you're confident you can solve. I outline in Chapter 25 how to break the problems down into simple, bite-sized pieces. Nothing is worse than sitting in an exam and struggling with a difficult problem for far too long. If you spend too much time struggling, you may miss really quick and easy point-getters later in the exam. Also, if you find and solve the easy problems first, you build up some momentum and confidence as you proceed to the hard problems.

State Your Assumptions

Start a problem by listing a few of the necessary assumptions. One or two words usually suffice. Check out the following example; you answer the questions and then jot down your answer in the test margin.

✔ Does this problem need to consider self weight? Is mass or weight mentioned somewhere on the picture or in the problem statement?

✔ Does this problem need to consider friction?

✔ What type of problem is this?

After you've listed your assumptions, seeing that list should help you identify the specific technique(s) you need to solve your problem.

Relax and Remember Your Basic Steps

Every statics problem, regardless of the type of problem, usually has the same basic beginning steps that never change.

1. **Draw a free-body diagram.**

2. **Write the equations of equilibrium for the entire system to find unknown support reactions.**

3. **Proceed to a specific solution technique based on your problem type.**

Even if your free-body diagram isn't totally correct, your instructor at least knows that you knew enough to start by drawing a picture and may award partial credit. Check out Part IV for more on drawing F.B.D.s.

Identify Your Origin and Coordinate System

As you're drawing your free-body diagram, make sure to clearly indicate your origin and Cartesian coordinate system. If you're working a three-dimensional problem, remember that you also need to include a z-axis. And don't forget to apply the right-hand rule (thumb is the x-axis, forefinger is the y-axis, and middle finger is the z-axis) so that you get the Cartesian axes properly oriented. Chapter 5 gives you the lowdown on mastering all things Cartesian.

Make sure that you clearly locate and display the origin with a dimension to each axis. You need that information when you start writing equations, especially if vectors are involved. If you're working a centroidal calculation type of problem, the origin is especially important because all of the centroid calculations you perform are based on relative dimensions (see Chapter 10).

Remember Your Vectors

Vectors are very useful for solving statics problems, both two-dimensional applications and three-dimensional problems. When working with vectors to ensure equilibrium, simply compute the *resultant vector* (a system of many similar effects transformed into a single equivalent vector) of all the applied forces and the resultant vector of the applied moments about a given point or axis. Set each of these resultants equal to $0\mathbf{i} + 0\mathbf{j} + 0\mathbf{k}$ and you have your equilibrium equations already computed. Head to Chapter 7 for more on resultants.

Write Your Equilibrium Equations

Nothing upsets a statics instructor more than not seeing an attempt at a free-body diagram, unless it's not seeing the three equations of equilibrium for two-dimensional problems written on the paper. Even if you write nothing else, put the following equations on a separate line for each problem:

$$\sum \left| \vec{F_x} \right| = 0 \quad \sum \left| \vec{F_y} \right| = 0 \quad \sum \left| \overline{M_z} \right| = 0$$

Writing these simple equations demonstrates that you at least understand the importance of the concept of equilibrium in the world of statics. And nothing makes a professor happier than knowing a student hasn't slept through all of his classes.

After you write these three basic formulas (or compute the resultants if you're working with vectors), use them as a guide for what you must do next, which is write each of these equations from the free-body diagram you (hopefully) drew earlier. (If you didn't draw it, get drawing!)

Stuck? Draw More Free-Body Diagrams

Depending on the type of problem you're solving, you may be required to solve for internal forces. In fact, if you're dealing with an application-type problem, you can almost guarantee it wants you to find at least one internal force at some point in the problem.

 ✔ **Truss problem** (see Chapter 19): If the problem is a method of joints-type of problem, draw an F.B.D. of joints in the system. If it's a method of sections problem, slice the truss into two pieces and draw an F.B.D. of one of those pieces.

- ✔ **Submerged surface problem** (see Chapter 23): Draw an F.B.D. of the submerged object and then include the hydrostatic pressure as a horizontal, linearly distributed load, as well as another force from the vertical weight of the fluid.

- ✔ **Frame/machine problem** (see Chapter 21): Break apart the system at the hinges, remove any pulleys and cables, and separate any tools or other strange objects on the system. Each of these pieces gets its own separate F.B.D.

After you have the diagrams drawn, write the equilibrium equations for each, which should give you a clue as to what to solve for first.

Draw Your Shear and Moment Diagrams Correctly

A shear and moment diagram is practically a gimme on a test, especially if the instructor gives you the applied loads and corresponding support reactions. As you sketch these diagrams, remember the following:

- ✔ **Work your diagrams from the left end of the beam.** The methods I describe in Chapter 20, and in particular the sign conventions, are all based on working from the left.

- ✔ **Your diagrams must come back to a zero value at the end.** If the reactions are correct and your shear and moment diagram doesn't come back to zero when you're finished, you know without a doubt that you've made a mistake in your calculations somewhere. If you don't have time to go back and correct it, circle the discrepancy and leave a note for the instructor that says, "This diagram doesn't close to zero for some reason, but I know it should!" You may not get full credit, but that simple statement may be worth a couple of points.

- ✔ **Concentrated loads cause jumps in diagrams.** Concentrated forces cause jumps in shear diagrams, and concentrated moments cause jumps in moment diagrams.

- ✔ **The order of the functions increases as you move down through the graphs.** The order of the shear function is one order higher than the load for a given interval, and the order of the moment curve is one order higher than the shear. Remember that the slope of the moment curve is directly related to the value of the shear.

✔ **Pay attention to positive and negative areas.** A positive area under a load diagram causes the shear to increase (become more positive), and a positive area under a shear curve causes the moment to increase (become more positive).

✔ **Always include units on your graph.** Don't make the instructor guess the units. Historically, instructors are very bad guessers.

Assess Your Answers

After you've worked through a problem on the test, go back and ask yourself whether your answer(s) make logical sense. This strategy is known as applying *engineering judgment.* A little common sense can help prevent problems, so take a moment after every problem and consider whether your answer seems reasonable.

Acknowledge Mistakes and Don't Erase

If you realize that you made a serious blunder but don't have time to rework the entire problem, you may be better off with a page full of slightly incorrect work than a page with no work. If you can find where you made your error, quickly include the correction and then write a couple of words about what the mistake was, where you made it, and the effect it has on the remainder of your solution process. Even if you don't fully correct the entire problem, at least you've identified the mistake, and that's a big step in conveying your competency to your instructor.

Index

Notes

Business/Accounting & Bookkeeping

Bookkeeping For Dummies
978-0-7645-9848-7

eBay Business
All-in-One For Dummies,
2nd Edition
978-0-470-38536-4

Job Interviews
For Dummies,
3rd Edition
978-0-470-17748-8

Resumes For Dummies,
5th Edition
978-0-470-08037-5

Stock Investing
For Dummies,
3rd Edition
978-0-470-40114-9

Successful Time
Management
For Dummies
978-0-470-29034-7

Computer Hardware

BlackBerry For Dummies,
3rd Edition
978-0-470-45762-7

Computers For Seniors
For Dummies
978-0-470-24055-7

iPhone For Dummies,
2nd Edition
978-0-470-42342-4

Laptops For Dummies,
3rd Edition
978-0-470-27759-1

Macs For Dummies,
10th Edition
978-0-470-27817-8

Cooking & Entertaining

Cooking Basics
For Dummies,
3rd Edition
978-0-7645-7206-7

Wine For Dummies,
4th Edition
978-0-470-04579-4

Diet & Nutrition

Dieting For Dummies,
2nd Edition
978-0-7645-4149-0

Nutrition For Dummies,
4th Edition
978-0-471-79868-2

Weight Training
For Dummies,
3rd Edition
978-0-471-76845-6

Digital Photography

Digital Photography
For Dummies,
6th Edition
978-0-470-25074-7

Photoshop Elements 7
For Dummies
978-0-470-39700-8

Gardening

Gardening Basics
For Dummies
978-0-470-03749-2

Organic Gardening
For Dummies,
2nd Edition
978-0-470-43067-5

Green/Sustainable

Green Building
& Remodeling
For Dummies
978-0-470-17559-0

Green Cleaning
For Dummies
978-0-470-39106-8

Green IT For Dummies
978-0-470-38688-0

Health

Diabetes For Dummies,
3rd Edition
978-0-470-27086-8

Food Allergies
For Dummies
978-0-470-09584-3

Living Gluten-Free
For Dummies
978-0-471-77383-2

Hobbies/General

Chess For Dummies,
2nd Edition
978-0-7645-8404-6

Drawing For Dummies
978-0-7645-5476-6

Knitting For Dummies,
2nd Edition
978-0-470-28747-7

Organizing For Dummies
978-0-7645-5300-4

SuDoku For Dummies
978-0-470-01892-7

Home Improvement

Energy Efficient Homes
For Dummies
978-0-470-37602-7

Home Theater
For Dummies,
3rd Edition
978-0-470-41189-6

Living the Country Lifestyle
All-in-One For Dummies
978-0-470-43061-3

Solar Power Your Home
For Dummies
978-0-470-17569-9

Internet

Blogging For Dummies,
2nd Edition
978-0-470-23017-6

eBay For Dummies,
6th Edition
978-0-470-49741-8

Facebook For Dummies
978-0-470-26273-3

Google Blogger
For Dummies
978-0-470-40742-4

Web Marketing
For Dummies,
2nd Edition
978-0-470-37181-7

WordPress For Dummies,
2nd Edition
978-0-470-40296-2

Language & Foreign Language

French For Dummies
978-0-7645-5193-2

Italian Phrases
For Dummies
978-0-7645-7203-6

Spanish For Dummies
978-0-7645-5194-9

Spanish For Dummies,
Audio Set
978-0-470-09585-0

Macintosh

Mac OS X Snow Leopard
For Dummies
978-0-470-43543-4

Math & Science

Algebra I For Dummies,
2nd Edition
978-0-470-55964-2

Biology For Dummies
978-0-7645-5326-4

Calculus For Dummies
978-0-7645-2498-1

Chemistry For Dummies
978-0-7645-5430-8

Microsoft Office

Excel 2007 For Dummies
978-0-470-03737-9

Office 2007 All-in-One
Desk Reference
For Dummies
978-0-471-78279-7

Music

Guitar For Dummies,
2nd Edition
978-0-7645-9904-0

iPod & iTunes
For Dummies,
6th Edition
978-0-470-39062-7

Piano Exercises
For Dummies
978-0-470-38765-8

Parenting & Education

Parenting For Dummies,
2nd Edition
978-0-7645-5418-6

Type 1 Diabetes
For Dummies
978-0-470-17811-9

Pets

Cats For Dummies,
2nd Edition
978-0-7645-5275-5

Dog Training For Dummies,
2nd Edition
978-0-7645-8418-3

Puppies For Dummies,
2nd Edition
978-0-470-03717-1

Religion & Inspiration

The Bible For Dummies
978-0-7645-5296-0

Catholicism For Dummies
978-0-7645-5391-2

Women in the Bible
For Dummies
978-0-7645-8475-6

Self-Help & Relationship

Anger Management
For Dummies
978-0-470-03715-7

Overcoming Anxiety
For Dummies
978-0-7645-5447-6

Sports

Baseball For Dummies,
3rd Edition
978-0-7645-7537-2

Basketball For Dummies,
2nd Edition
978-0-7645-5248-9

Golf For Dummies,
3rd Edition
978-0-471-76871-5

Web Development

Web Design All-in-One
For Dummies
978-0-470-41796-6

Windows Vista

Windows Vista
For Dummies
978-0-471-75421-3

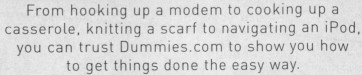